Motivation and Productivity in the Construction Industry

Motivation and Productivity in the Construction Industry

Robert H. Warren

VNR Van Nostrand Reinhold
New York

Softcover reprint of the hardcover 1st edition 1989

Library of Congress Catalog Card Number 88-27451

ISBN 978-1-4684-8829-6

Printed in the United States of America

Text Design by Sandra G. Cohen

Van Nostrand Reinhold
115 Fifth Avenue
New York, New York 10003

Van Nostrand Reinhold (International) Limited
11 New Fetter Lane
London EC4P 4EE, England

Van Nostrand Reinhold
480 La Trobe Street
Melbourne, Victoria 3000, Australia

Macmillan of Canada
Division of Canada Publishing Corporation
164 Commander Boulevard
Agincourt, Ontario M1S 3C7, Canada

2 4 6 8 10 12 14 16 15 13 11 9 7 5 3 1

Library of Congress Cataloging in Publication Data

Warren, Robert H.
Motivation and productivity in the construction industry / Robert H. Warren.
p. cm.
Includes index.
ISBN 978-1-4684-8829-6 ISBN 978-1-4684-8827-2 (eBook)
DOI 10.1007/978-1-4684-8827-2
1. Construction industry—Personnel management. I. Title.
HD9715.A2W34 1989
624'.068'3—dc19 88-27451
CIP

To the men and women of the construction industry who build the things that make life possible and living enjoyable.

Contents

Foreword

Bob Warren has given us a construction management book with a mission: to ensure that the "People Quotient" is factored into all facets of project management and job-site performance.

Warren demonstrates the importance of respect for every craftsman's individual worth and of empathy for the conditions surrounding construction employment.

He outlines the motivational forces that achieve highest productivity and, at the same time, displays on every page a broad generosity of spirit along with keen analytical powers.

Clearly, the book is the fruit of the author's own years of experience in the field. I am proud to say that 13 of those years were spent with Fluor Daniel, so that I can vouch personally for his acumen and leadership in project management.

I commend this book most heartily to my colleagues, clients, and competitors throughout the construction industry. It should be on the shelf of every caring manager who wants, not only to get the best out of his workforce, but also to accomplish the best for the members of that work force as individual human beings.

Les McCraw, President
Fluor Corporation
Irvine, California

Preface

Today we frequently hear that construction productivity has slipped because labor "just isn't as good as it was in the old days." The truth is that labor as a resource is better than it was in those "old days." In reality, what has slipped is the way the resource is managed. Managers simply have not kept up with the changing nature of the work force.

This book is intended to provide the reader with practical guidance in developing a motivational climate for the construction craftsman that will permit productivity to increase and overall project costs to decrease.

The relationship between motivation and productivity is such that improvement in one causes an improvement in the other. A self-propagating effect takes place. This concept is explored in detail, and solutions on how to improve productivity through application of motivational techniques are described.

When specific action is taken to improve productivity by bringing the plan, materials, tools, and labor to the work site at the right time and in the correct sequence, an improved climate for motivation of the craftsman is created. When specific action is taken to enhance motivation through establishment of meaningful communication, consistent management, removal of frustrations, fine tuning of the work force, maintenance of a safe environment, attention to basic personal comforts, training and recognition of achievement, productivity will improve. These subject areas are addressed, and the interweaving of each with the others is analyzed.

To help the reader apply the material presented in this book to actual situations, the last chapter contains an outline of how to recognize the signs and symptoms that indicate productivity is suffering due to poor motivational levels in the craft work force, how to implement a motivation-improvement program, and how to gauge the ongoing efficacy of such a program.

Experience gained by the author during an engineering and construction career that spans over thirty-five years has been called upon to help the novice manager understand how the interaction between motivation and productivity can separate a good construction job from a bad one. Another goal is to assist the seasoned manager who recognizes that improved human relations skills are needed if one is to stay on the leading edge of cost performance in today's competitive markets.

Some anecdotal material has been included to help others, including human resource managers and instructors in construction management, relate some of the more abstract concepts of the engineering and construction project to common everyday experiences.

Finally, for the project manager who is confronted daily with the challenge of bringing the project in on time and under budget, the typical craftsman's thoughts and reactions to the multitude of stimuli on the construction job have been verbalized. The hope is that, so informed, the project manager will be able to create and maintain a proper motivational climate.

In our quest for productivity improvement, we have subjected practically everything under the sun to the heat and pressure of technological change. Now it is time for a metamorphosis in the way we manage the craftsman.

Acknowledgments

This book would not be complete without acknowledgment to the several thousand craftsmen, supervisors, managers, engineers, administrators, and clerical people who, although none of us knew it at the time, contributed to the material in this book. Our constant association over several years provided the basis for the anecdotal material; their words, deeds, actions, and reactions created the fertile ground from which the fundamental concepts grew.

A special thanks is due to Les McCraw for his counsel, especially for emphasizing the role of a safe working environment as a strong motivator. Thanks also to George Fischer for our countless "rap sessions" while the manuscript was being developed. And, to Todd O'Donnell, who many years ago taught me the meaning of empathy.

In addition, I am grateful to Buck Mickel for propagating the late Charlie Daniel's craftsman-oriented philosophy, a philosophy in consonance with the basic relationship between motivation and productivity as it is presented herein.

And finally to Rob, Richard, Elizabeth, and Paulette who, through their interest and encouragement during some unsettled times, helped keep the end goal in sight.

1

Concepts and the Positive Motivational Climate

The thrust of this book is more pragmatic than philosophical, more empirical than theoretical. It has been written in practical terms and in layman's language so that the reader can focus on the essence of its message: the relationship between a craftsman's motivational level and his productivity.

Although for simplicity in presenting the material in this book, references to craft labor are made in the male gender, it should be emphasized that over 2 percent of craftsmen in construction today are women and that the content of this book applies equally to them. Women have been joining the ranks of skilled craftsmen since the mid-seventies in a variety of crafts. Today they work as welders, instrument fitters, electricians, insulators, and equipment operators, to mention a few. Training-class instructors indicate that women acquire skills involving hand-eye coordination very quickly. This is especially important when the tasks involved must be finished and neat in appearance as well as functional, as is true of many exposed-insulation installations, for example. In addition, women are extremely adept at operating machinery designed to close tolerances. On one project, the operator of the primary piece of loading equipment—a

Caterpillar Model 988 that had a 15-cubic-yard bucket, was 33 feet long and 15 feet tall, weighed 50 tons, and was the largest rubber tire end loader manufactured in the United States—was a young woman barely over 5 feet tall. The vertical distance to the cab, accessed by a ladder, was over twice her height. Her superior dexterity skills in operating the hydraulic power assists enabled her to make vertical and horizontal cuts that required little hand work when she finished.

Webster defines *productivity* as the act of producing in abundance; the term as used herein means the completion of construction work at unit rates more economical than the average, less than those published in estimating handbooks, and better than those used in producing the estimate for a given project. Similarly, *motivation* is characterized as a combination of influences that causes the craftsman to want to do the job as quickly as possible consistent with safety and quality goals while cooperating, on a larger scale, with his fellow craftsmen in execution of the project as a whole.

PRODUCTIVITY

When a manager says that productivity was good on this job and bad on that or good in this area and bad in that, the reference is usually to how costs on that job tracked in comparison with either the estimate for the project or the unit man-hours usually allotted to that particular item of work, based on company historical cost data.

The advent of the jet age and instant telecommunication capability across the United States allowed many contractors to pursue work from coast to coast and border to border. Historically, as they branched out, they used their hometown productivity units as a base and applied up or down factors for the new areas. Many Gulf Coast companies, for instance, used production rates they experienced in that area during the 1962–1963 period, with appropriate location factors, for figuring work in Nebraska, Colorado, Washington, and other distant areas. Natural cross-pollination of estimating personnel caused these base rates to become the standard for many firms in the east and southeast also. One major company, in developing its construction estimating manual, based all its productivity rates on its experience over the years in the midwestern city that was its home base and predicted productivity factors for other parts of the country with modifications. Regardless of their origin, the estimating factors that can be used after implementation of a motivation program will be more productive than those in use before it. Good productivity then will be considered attained when these units are bettered on a given real project.

MOTIVATION

Motivation as a concept is somewhat more abstract than productivity. Many industrial psychologists maintain that a manager can only create a climate; craftsmen must motivate themselves. As used herein, motivation is characterized as the group of influences (climate being one of them) that cause craftsmen to *want* to perform a given task. Motivation to do something is present in everyone, to some degree, all the time.

As a simple illustration, a small child wants to eat his dessert rather than his spinach. He gets more pleasure from the taste of the former, more basic satisfaction. He is *motivated* to eat his dessert. On a somewhat higher plane, motivation causes the figure skater to spend countless hours going over and over routines to perfect them. She knows that practice makes perfect and that she cannot excel in her next competition unless her execution is flawless. She wants the ego satisfaction that comes from accomplishing her goal. She is *motivated* to practice.

INTERDEPENDENT RELATIONSHIP

For many years, behavioral specialists have been saying that in an industrial climate, increased motivation causes increased productivity; that productivity is dependent upon motivation. This is patently true, but it is only half the story. The other half is that increased productivity causes increased motivation, that motivation is in return dependent upon productivity. Figure 1-1 illustrates this interdependent relationship. In the figure, as motivation is increased from a_1 to a_2, productivity increases

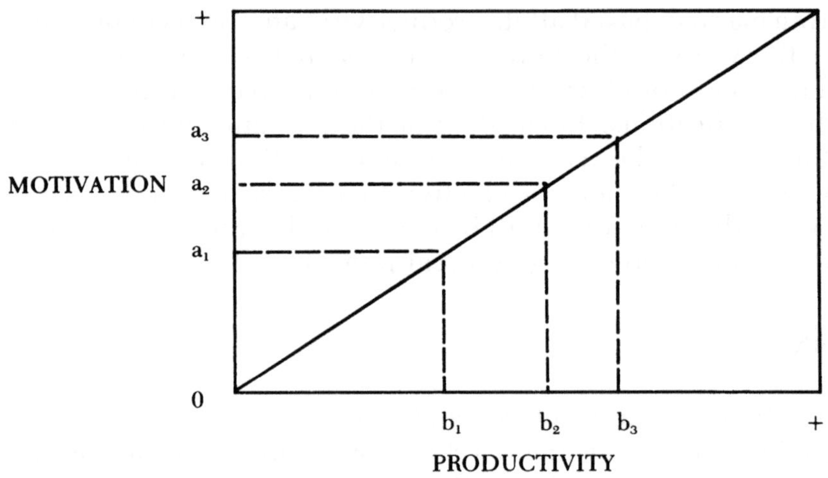

Figure 1-1 Motivation vs. Productivity

from b_1 to b_2. Conversely, when productivity is increased from b_1 to b_2, motivation will increase from a_1 to a_2.

This may, at first, seem a contradiction, but in the context of finite incremental changes, experience shows that an increase in one causes a like increase in the other, and, similarly, a decrease in one causes a decrease in the other. They are interdependent. This major point will be illustrated in many different lights.

EGO SATISFACTION

As a team or crew member, ego satisfaction is what causes the craftsman to strive to do his part efficiently and correctly the first time. Productivity increases as a result. The manager and the craftsman are thus working for the same thing: the elimination of waste motion, errors, and miscoordination. The manager does it out of a desire to reduce costs and allow the company to remain profitable; that, in essence, is his job. The craftsman wants efficiency because of the personal satisfaction that comes from being able to perform his craft without delays and other interference. A self-propagating relationship between motivation and productivity is thus generated and, when properly fed, intensifies with each repetition until the maximum achievable results are obtained. This can be illustrated by referring again to Figure 1-1. Experience indicates that as motivation goes from a_1 to a_2, productivity can go not only from b_1 to b_2, but beyond to b_3, thus in turn, producing a further increase in motivation from a_2 to a_3.

When management brings the plan, the materials and machinery, the equipment and tools, and the labor to the point on the site at which the work is to be installed, at the correct time and in a sequence that allows the craftsman to do the job efficiently, two results are produced. First, more work is produced in a given unit of time and costs go down. Second, the craftsman produces more than an improvement in work flow alone would otherwise stimulate. When he can work efficiently, without the frustrations brought on by poorly coordinated work flow, he derives more satisfaction. He enjoys doing the job he was trained for. He therefore motivates himself to produce more because it is a rewarding experience for him.

This is an emerging concept in the way construction labor is managed, and it is in sharp contrast to practices of the past. A closer look at the way in which labor used to be managed will illustrate this concept.

THE "CUSSIN'" APPROACH TO LABOR MANAGEMENT

The most prevalent method of managing construction labor well into the 1950s and, in many cases, the 1960s was called by many the "cussin'" approach. It was also known more formally as "management by threat." The general idea was that the more you yelled and cursed at an employee, the harder the employee would work. Naturally, those who yelled the loudest and cursed the longest climbed rapidly up the supervision ladder. The consummate superintendent was the one who could curse between the syllables of his words.

A small boy was first exposed to this approach in the construction industry when, one morning, he watched a group of workers in front of his house. Some had shovels and others had picks, and they were all being addressed loudly by a man in a large straw hat (the rest had caps on) who waved his arms wildly and pointed in various directions, frequently using words that were unknown to the child. The workers lined up a few feet apart and began digging along the side of the street. The man in the straw hat retired to the shade of a tree. Occasionally, he went to one of the workers, leaned very close to him, and, with hands on hips or pointing excitedly to the ground, yelled out a stream of those unfamiliar words. The boy continued to observe this curious scene until his mother eventually heard the commotion and the language and took him away from his vantage point.

This occurred during the great depression, and what the boy had seen was a WPA (Works Progress Administration) project in which the gutter at the side of the pavement was being replaced with a granite stone curb. The strange words were profanity, and what he witnessed was the then prevailing technique of managing the laborer, the "cussin'" approach: if you berate, threaten, intimidate, and tell the worker nothing except where to dig, you'll get the maximum amount of work done.

Many years later, as a new engineer fresh out of college, the young man quickly recognized this approach again on his first construction project; even the big straw hats versus the caps were the same, but he now realized that the headgear was a status symbol designed to distinguish the boss from the worker. The "cussin'" approach had not only survived the great depression and World War II; it had actually flourished.

There were two primary reasons for this. First, the work force was made up of veterans—either of the depression or of the war; in many cases, of both. The military excelled at the "cussin'" approach, could even have invented it. These supervisors knew no other style of management. Second, and really more significant, the craftsman had little choice but to take the abuse. Jobs were scarce, and it was an easy decision for a craftsman to grit his teeth and bear the mistreatment rather than stand up against it and risk being fired.

The main goal of the craftsman in those days was, literally, survival: earn enough to put clothes on the children, a dry roof over their heads, and food on the table. He would produce just enough quantity and quality to get by and avoid the wrath of the supervisor. Sometimes, the more inventive craftsmen would devise ways to conserve energy (theirs) by making themselves look busy while actually coasting. This was accurately called *busywork*.

For example, someone using a pick would exert energy only to lift the pick, letting it fall of its own weight on the downstroke. The sheet-metal worker similarly could generate a cacophony with a hammer while producing little usable duct work.

The craftsman's motivational level was virtually nil. If anything, the motivation was only to survive until the quitting whistle signaled the end of the day and release from the job. The worker was tuned out and turned off to management and management's needs primarily because of low aspirations and an even lower self-image. He believed what he was constantly hearing—"weakling," "stupid," "lazy," "clumsy"—from his supervisors. And management did not provide any sort of climate in which he could satisfy his growth needs even if he had any.

CHANGING ATTITUDES

A number of influences during the late 1950s and early 1960s changed attitudes in American industry in general and in construction especially. An evolutionary transition in the makeup of the work force began. For example, average educational levels rose. Craftsmen had to become more educated simply to keep up with rapidly increasing advances in construction technology. Workers also began to aspire to higher goals in life. Gradually, they started to want a better life for their families and for themselves; they wanted to satisfy something beyond their basic survival needs. Television influenced this change in attitude greatly, perhaps more than any other factor. Through it, workers became aware of a whole new set of values, standards, and statuses. They were led to believe by what they saw in television programming and advertising, that they could rise above their present stations. Their concept of self-worth improved, notwithstanding the influence of their supervisors.

Later chapters will describe this new generation of craftsmen and will illustrate the collection of influences and factors that caused them to want to produce results with their labor. Another important fact will also be analyzed: the worker is excited or disenchanted by many of the same factors as is the manager or supervisor.

Labor was not the only group to undergo change during the late 1950s and early 1960s. Managers' styles started departing from the old stereotypes described above and became more enlightened. Through better training, young managers began to gain an awareness of the psychological needs of craftsmen. Much of this awareness was brought about by the cross-pollination effect created by the natural movement of managers, supervisors, industrial psychologists, and similarly trained personnel between the manufacturing and construction industries. Typically, this transfer of ideas and information would occur when one of these people, having been exposed to new techniques through employment in and training received in manufacturing, would move to construction in a career advancement.

Manufacturing had begun to recognize, even in the early 1950s, that productivity in the United States had to improve to keep pace with postwar competition developing in world markets. There was, therefore, a great deal of study and experimentation in the behavioral sciences in an attempt to bring about that improvement. Construction was able to benefit—get a "free ride"—from the emergence of revolutionary theories that were first tried and tested in the manufacturing industry.

MASLOW'S HIERARCHY OF NEEDS

One such theory was developed by Abraham H. Maslow, a noted behavioral scientist and chairman of the department of psychology at Brandeis University. On the basis of a basic motivation theory he conceived and presented to the psychoanalytic profession in 1942, he published, in 1954, his hierarchy-of-needs theory that interpreted human behavior in terms of psychological needs (*Motivation and Personality*, Harper & Row, Publishers, 1954).

This theory suggests that human beings are driven by a series of psychological needs that manifest themselves as goals. The needs start, Maslow submitted, with the most basic and escalate up a scale of necessity and intellectual significance to the highest (or perfect) intellectual state, reflecting self-fulfillment. Further, he explained, when a particular lower set of needs is satisfied, the individual becomes psychologically restless and strives for the next higher set. The former goal is then put in a sort of successful-accomplishment category and thereafter taken for granted, essentially considered by the individual as a human right.

Maslow identified five levels in his hierarchy (in ascending order).

1. physiological needs
2. safety needs
3. belongingness and love needs
4. esteem needs
5. self-actualization needs

He also postulated a set of conditions which themselves are prerequisites to the hierarchy needs. These preconditions—which are, in fact, considered rights in a society such as ours—include freedom of speech, freedom to do as one wishes within the law, and freedom of expression as well as certain other cultural qualities, such as justice, fairness, honesty, and decorum. Threats to any of these prerequisites are reacted to by the individual with emergency response actions designed to defeat the threat to both the prerequisite and the underlying basic need.

It is important to recognize that the dividing lines between the various levels of needs are somewhat indistinct. At any point in time, a person may drift from one level to another, either for a period of time or with regard to a specific category. Once having achieved one of the levels in the hierarchy, an individual's more noticeable departures from it are usually downward on the scale. For example, a craftsman who has recently lost some sense of security as a result of being discharged may be temporarily dislocated from the love-and-belongingness needs level to the safety needs level until he again becomes employed. In construction, he will probably encounter this change many times in his career because the employers for whom he will work hire and lay off workers as their work load fluctuates.

Physiological Needs

The needs on this level may be characterized as those necessary for survival of the body. They include the need for food, water, shelter, warmth, and sleep. For example, an individual dominated by needs at this level might be totally consumed by a chain of subgoals: the need to earn money in order to purchase food in order to satisfy hunger (a basic survival need). The desire to work in a cleaner environment or perform more intellectually challenging tasks may exist, but it is subordinated to the individual's interest in preserving the food supply.

In many cases, threats to physical safety and personal well-being may be ignored in the quest for survival. Organizational behavior is dictated by these needs. The individual will tolerate almost any adversity in connection with the job in order to achieve or maintain the certainty of eating. Once this certainty is achieved, however, the motivating influences immediately evolve to the next higher level—safety.

Safety Needs

At this level, the individual's concerns are for safety and the causes that ensure or jeopardize it. Characteristic of these needs are the desires for security; for stability in relationships with others; for freedom from harm, fear, anxiety, and chaos; for the presence of a strong protector; for a structured environment; for predictability of events; and for law and order. These needs are not necessarily physical in character. For example, a strapping 6-foot-tall ironworker may never feel the need for a strong protector in a physical sense, but he may need, very urgently, the sheltering of a strong protector in the form of a superintendent who can add security to his tenure of employment.

The individual at this level will show preference for an undisrupted routine or rhythm. Unanticipated changes in plan and surprises will be upsetting and cause loss of efficiency. There will be negative reactions to outbursts and harsh treatment from the supervisor. The worker will not perform well in new situations and unfamiliar surroundings and may become inwardly terrified when confronted by disorganized, unmanageable, or chaotic events or by the loss of things formerly depended upon.

But when convinced that the environment is relatively safe and unthreatening and that he is not in harm's way, the worker aspires to the goals of the next-higher level of needs. When survival needs—physiological and safety—are fulfilled, they no longer serve as motivators. The individual now seeks a sense of belongingness and love.

Belongingness and Love Needs

When the individual was, as above, concerned with where the next meal was coming from, where he would sleep at night, or whether he would be fired from his job, love and belonging were unimportant to him. His preoccupation was with survival. With those needs satisfied, however, he will search out relationships that provide him with these higher needs. They are social in nature in comparison to the lower ones, which are essentially survival-oriented. He will seek love and affection in his family. Again, using our ironworker as an illustration, his need for love may not be characterized by an open display of affection but by a search for the friendship of his fellow craftsmen. He will attempt to avoid loneliness, ostracism, and rejection. He will strive to achieve a state of acceptance and a feeling of belonging to a group or organization. He will gather after work with his friends for a soft drink or glass of beer. He will reinforce his perception of belonging, once achieved, by demonstrating the relationship openly. An employee confirming association with his employer might wear a baseball cap or jacket with the employer's name and logo emblazoned on it. A newly appointed supervisor might display a somewhat more subtle symbol by carrying an array of pens and pencils in his shirt pocket to announce that his promotion has placed him in a new group, supervision, in which performing paperwork is perceived by the worker as a responsibility.

A particularly competent supervisor who had, during a twenty-five-year career, progressed up through the construction-craft ranks from pipe-fitter helper to general superintendent was considered for promotion to project manager. The project was a contract maintenance and expansion program that he had, because of his familiarity with the work, already been in day-to-day charge of for several years. His excellent performance on the job had convinced both the contractor and the client that he would make a successful transition from superintendent to manager, from the number-two slot in the project organization to the number-one slot. Accordingly, he was promoted, and the previous project manager was transferred. No other changes were necessary because the project was running smoothly. A few weeks later, the new project manager's boss visited the project and was greatly surprised to see him, not in the khaki work clothes that for decades have been the choice of supervisors in the construction industry but, for the first time ever, outfitted in a new white shirt and tie. The project manager's proud way of illustrating his belonging to a new group, management compared to supervisory, was to don what he perceived to be the uniform of the office.

As with the lower-level needs, when the individual feels that he has accomplished the goal of acceptance, belonging, and being loved, he strives for the next level of needs, esteem.

Esteem Needs

Esteem has two components: self-esteem and the esteem of others. A person seeking self-esteem exhibits a desire for the achievement and mastery of some quality or skill. He or she strives with intensity to develop competence in some area of performance. For the craftsman, this desire manifests itself in pride of accomplishment: the welder is given two pieces of pipe to weld together and gains a sense of self-satisfaction from doing so while laying down a perfect bead of molten metal; the carpenter is told to build a set of concrete forms to certain dimensions, and when the forms are stripped off, he is satisfied if the concrete is exactly as he planned it would be.

Self-esteem enables a person to develop a sense of self-worth, adequacy, and usefulness that permits him to face the world confidently. If these feelings are thwarted, he will feel discouraged, inferior, and weak and regress to the earlier level of needs, the need for love and a sense of belonging.

Self-esteem develops, to some extent, from the esteem of others; that is, what others think. The welder first has to be shown by the instructor what a perfect weld looks like and then receive praise and recognition for making one. Learning how to make a perfect weld will certainly increase the welder's self-esteem, but the desire for the esteem of others may continue to make the approval and praise of the authority figure—in this example, the instructor—a sought-after element. The esteem of others is therefore important not only in setting standards of performance but also in fulfilling a desire for reputation and prestige. The welder believes he is performing the operation correctly, but approval and praise from someone else is needed to reinforce this belief.

Maslow cautions against basing one's self-esteem too much on the opinions of others without realistic consideration of one's true capacity, accomplishment, and adequacy of performance. He suggests that the healthiest external influences on self-esteem are those based on the bona fide esteem of others; esteem that the individual, being honest with himself, knows is well-deserved.

When the individual feels confident that he has the esteem of others, he seeks self-actualization, the highest level of motivation.

Self-Actualization Needs

A drive toward self-actualization means that the individual, being true to his own nature, is motivated to become what he feels he must be. He wants to become what he believes he is best suited to be and all that he is capable of becoming: He wants to make real and actual the potential that he sees within himself. He pursues his ultimate goal: self-fulfillment. If his dream is to be one of the best welders, that is what he strives to be; if he wants, instead, to be a general superintendent on a large project, that is the goal he pursues. He attains self-actualization when he "actualizes" whatever he set out for himself as an ultimate accomplishment.

Once in the self-actualized state, his motivation comes from his desire to continue doing what he knows he does best. He performs iteration after iteration of the tasks involved and obtains satisfaction each time.

McGREGOR'S THEORIES

In 1960, Douglas McGregor of the School of Industrial Management at Massachusetts Institute of Technology introduced his now classic Theory X and Theory Y concepts that described two completely divergent views of labor and their related management styles for directing human energy. He showed convincingly that, rather than being essentially indolent and lacking in ambition, the worker has, indeed, the capacity and desire for organizational involvement and the readiness to assume responsibility. All that management has to do to unlock the potential, he offered, is to recognize the human side of enterprise and act accordingly.

In order to provide clarity in discussion and to avoid clouding with potentially complicating labels the basic thrust of the two propositions he compared, he simply called one *Theory X* and the other *Theory Y*.

Theory X

McGregor described Theory X as the conventional idea of the way in which management uses human energy to meet its organizational needs. Not unlike the "cussin' " approach described earlier, Theory X has three elements.

1. Management is responsible for organizing people, materials, and machines to achieve its economic ends.
2. This responsibility involves directing people's efforts, motivating them, controlling their actions, and modifying their behavior to fit the needs of the organization.
3. Unless people are so managed, they will be passive or, worse, resistant to the needs of the organization. Management must therefore closely control their activities through persuasion, reward, and punishment.

Theory X further characterizes people in the following way:

- Workers are lazy, lack ambition, and work as little as possible.
- They dislike responsibility and prefer to be led.
- They are self-centered and indifferent to organizational needs.
- They are resistant to change.
- They are dim-witted, gullible, and easily duped.

Theory X further postulates that management, faced with the above human fiber with which to accomplish work, has two choices. The first is to follow a hard line, using tight controls, coercion, and threats. The second is to take a soft approach characterized by permissiveness and accession to workers' demands in order to make them tractable and more likely to accept direction.

He then explains the inherent defects that make each of these approaches wasteful of human energy and therefore inefficient. The hard line results in antagonism and confrontation, militant unionism, restriction of output, and sabotage of management's objectives. The soft approach leads to abdication of management control, indifferent performance, and higher costs for lower output. Neither approach results in improved productivity; neither creates a climate conducive to motivation of the individual.

McGregor then suggests that the cause of human behavior problems in industrial organizations is the manner in which workers are managed, not the workers themselves. He says that the conventional approach represented by Theory X is inadequate because it is based on a mistaken notion of what is cause and what is effect. The answer, he postulates, is Theory Y.

Theory Y

Theory Y, says McGregor, provides an alternative to the conventional management concepts of Theory X. Theory Y is based on more adequate assumptions regarding human nature and human motivation. It has four elements, the first of which is identical to the first element in Theory X.

1. Management is responsible for organizing people, materials, and machines to achieve economic ends.
2. People are not naturally passive or resistant to organizational needs; poor management has made them that way.
3. All people possess latent motivational capabilities, the potential for development, the capacity to take on responsibility, and the readiness to work toward organizational goals. Management's task is to nurture these tendencies and help people develop these characteristics in themselves.
4. Management should structure its organization and its methods of operation so as to allow people to set their own goals and direct their own efforts toward the organization's objectives.

Management can do this by creating opportunities, removing obstacles, encouraging growth, and providing guidance, McGregor maintains. In essence, Theory Y recognizes the contribution workers can make toward the organization's goals, if management will allow them to do so. This means, however, treating workers as mature adults instead of children, minimizing external controls, and relying on the workers' ability to direct themselves along the same footpath as management. It does not lead to the abdication of management control, elimination of leadership, or lowering of expectations that might be experienced in the soft approach of Theory X. On the contrary, it involves unleashing the potential present in all people and guiding it toward achievement of the organization's goals.

McGregor maintained that implementation of Theory Y would result in substantial improvement in the effectiveness of the industrial organization. To begin the transition from the old styles (Theory X and its derivatives) prevalent at that time, he suggested several steps management could take.

- Delegation: Give people a degree of freedom to assume responsibility in helping to achieve organizational goals, to direct their own activities, and to gain ego satisfaction at the same time.

- Job Enlargement and Participation: Encourage people at all levels of the organization, including the lowest levels, to assume responsibility, to direct their creative energies toward organizational goals, and to participate in developing solutions to problems. Give them a voice in shaping the decisions that affect them, and provide opportunities for ego satisfaction.
- Goal Setting and Performance Appraisal: Involve individuals in setting goals of their own that will act in concert with those of the organization. Encourage them to take greater responsibility for planning and appraising their contribution to the organization's objectives. Involve them in self-appraisal of their performance along with the performance of their supervisor. Praise achievement, recognize when improvement is required, and spell out specific actions which can become new goals.

As did Maslow, McGregor raised some cautions to implementation of his theory. He predicted failure if management subscribed to the concept of Theory Y but applied the concept within the framework of Theory X. He also saw lack of sincerity in its application as preventing success. If the concept were applied as a sales gimmick or as a device for tricking people into thinking management viewed them as important when it really did not, failure was certain. If the theory is to be grasped, he felt, management has to have confidence in human capacities and must also direct itself toward organizational objectives rather than preservation of personal power.

The combined impact of Maslow and McGregor on the worker-management relationship is enlightenment. The sharp contrast between the "before" and "after" cases is illustrated in Figure 1-2, which combines the key terms of both these concepts in a single matrix.

EXPECTANCY THEORY

In the late 1960s, still another concept, the Expectancy Theory, was developed to predict the relationship between moti-

ELEMENT	BEFORE	AFTER
Management Labor	Theory X Survival	Theory Y Satisfaction

Figure 1-2 Changing Attitudes

vation and performance in worker behavior. It suggests that an individual's level of motivation in performance of a task depends on three perceptions.

1. How he rates the attractiveness of the result that an improvement in his performance will have.
2. How he estimates the probability that increased effort will lead to better performance.
3. How he perceives the probability that better performance will lead to the desired result.

The theory emphasizes that in all cases, it is *perceived* probabilities that govern the worker's actions, not actual results; that is, it is what he *expects* will occur, not what actually *does* occur.

For example, he may consider a raise in his wages as an attractive goal. While he may believe that an increase in his effort will lead to improved performance, he may not believe that improved performance will necessarily lead to a pay raise. Whether it does or does not, his motivation depends on what he expects would happen in the situation.

If, on the other hand, the result he strives for is pride in workmanship and sense of accomplishment, he may expect that increased effort will lead to improved performance, which will, in turn, lead to that reward. He will thus be motivated to increase his effort.

Not all results are perceived as attractive. Suppose that the worker perceives that the outcome of his extra effort will be improved performance but thinks that improved performance will result in exhaustion. Whether it will or will not, his perception is that it will; and so he will not be motivated to expend the extra effort.

The key point of the Expectancy Theory is that the more attractive the reward and the stronger the perception that changes in effort will produce like changes in performance and that improved performance will lead to the reward, the higher the motivation level will be. This theory, then, can be used to predict the motivational consequences of not only pay changes, but also promotions, changes in working conditions and assignments, use of overtime, training, and recognition of achievement, to name but a few.

The distillate of the Expectancy Theory as it applies to the construction industry is that even the lowest-paid craftsmen, by virtue of their socioeconomic and psychological development, have needs and expectations today that go beyond what they were perceived by management to have just twenty-five or thirty years ago. And for craftsmen to be motivated, they must at least see a potential for fulfillment of these higher needs and expectations.

EMPATHY

Recognition of these facts by managers is an important first step in developing the ability to manage today's craftsmen. Managers must put themselves mentally into the craftsman's place and understand his feelings. This process is called *empathy*: the ability to project one's self into another's feelings and ideas.

Younger managers began to take heed of these theories in the 1960s and 1970s and abandoned the rough, gruff "cussin'" style they have been exposed to in prior years. They found that by skillfully using empathetic reasoning, they could predict accurately the craftsman's reaction to virtually any stimulus or, more important, antistimulus. By simply asking themselves what they would want or what they would do were they in the craftsman's shoes, they could anticipate the craftsman's actions.

For example, before changing the time of day the work force is paid from early morning to late afternoon, these younger managers would give a thought to how many of the employees needed to dash to the bank with their paychecks during their lunch hours in order to satisfy an obligation and how, if the time were changed to after lunch, they might be delayed an additional day in depositing their paychecks. The truth is that managers also frequently need to get to the bank but, unlike the employees, they can slip away during the workday. In reality, the needs of people in the working environment do not differ greatly from one group to the next. If the manager is paid twice what the craftsman is, the chances are that he, because of his lifestyle, also has twice the obligations on payday. The analogy goes further, and, in fact, one of the readily observable phenomena in today's society is that labor and management, craftsman and supervisor, have very similar needs and wants.

Today on television, for ten minutes out of every hour, we are barraged with advertisements for literally everything and anything money will buy. Managers see a particular commercial for a new car, appliance, or vacation in Europe, and they want it. Craftsmen see the same commercial, and they want the same thing. Due to the influence of television, our needs for both material and psychological growth are converging.

In 1960, a labor general foreman on a large highway and bridge project in the northeast took advantage of a two-week-long winter-weather shutdown and drove to Miami. He had seen the advertisements on television, prevalent at that time of year, extolling the warm sun and sandy beaches of south Florida as a getaway spot from the cold, frigid weather up north. He was single and had few obligations, so he just made a hotel reservation, got in his car, and went. Just as the bellhops were unloading his luggage at one of the finer ocean-side hotels, the big boss and owner of the company came out the front door. He, too, had been enticed by the thought of warm weather and had taken advantage of the same job shutdown to obtain relief from winter.

Unfortunately, in those days there was little understanding by management of empathy, and the big boss was surprised to the point of anger by the thought that a lower-level supervisor could have the same needs, wants, and aspirations as he, the owner of the company. He couldn't understand it and couldn't accept it. He could not emphathize.

His reaction to the labor foreman was typical of his entire management style. Because of it, he couldn't keep good people working for his company. New hires would quickly recognize the prevailing atmosphere and depart as soon as they could. As a result, the company had a virtual revolving-door personnel department; the only people who stayed were those who had difficulty locating other work.

The company eventually failed because the owner could not execute projects for the money he had in his estimates. Even though his prices were in line with those of other contractors bidding similar work, he could not get the productivity from his work force that the others did. He paid dearly for his demotivating climate.

Today's manager, because of better training in industrial psychology, is in the enviable position of being able to project the needs of the work force by simply looking at his own needs. The rules are the same. If he wants something for himself, the chances are his people want it, too; if there is something he would not do, he should not expect his people to do it. In a more positive sense, the manager who is able to visualize, through empathy, the set of conditions that will influence craftsmen to want to do a job—to look forward to coming to work in the morning—is the manager who will succeed in meeting or beating his cost estimate every time.

THE POSITIVE MOTIVATIONAL CLIMATE

How does the manager begin to create a positive motivational climate? First, he provides direct support to his craftsmen; second, he provides psychological nourishment to allow them to motivate themselves.

Direct Support

In providing direct support to the craftsmen, the manager is in the role of a coordinator. He must assure, through his direction to his subordinates, that all things necessary to complete a particular item of work will arrive at the point at which the work is to be carried out on schedule and in the correct sequence.

If, for instance, the assignment is to install a water pipeline, the manager should make sure that the survey party has staked out the alignment before the excavating equipment arrives; that the trench work has progressed sufficiently before installation starts; that the pipe, fittings, joint materials, and gaskets are convenient to the trench; that the back-fill material and the equipment to place and tamp it is available; that any required inspections or tests are provided for on a timely basis; and, finally, that the labor crew has been properly briefed with regard to specifications and safety hazards.

Though this may seem to be belaboring the obvious, experience shows that a surprising amount of the miscoordination on construction projects arises from similarly simple oversights.

The manager prepares his plan by thinking through the series of events and activities that must occur to progress the portion of the work at hand, hour by hour, day by day, or whatever increment he is comfortable with. He starts by imagining, or recalling from previous experience, the influences that will cause the activity to speed up and those that will cause it to slow down, those that will stimulate progress, those that will impede it. He then works out plans to exploit the positive influences and avoid the negatives. Success in this endeavor will depend on how well he can view the tasks from the craftsman's position. After all, the craftsman is the central figure in getting the work done.

A saying once used in the military to characterize the importance of the foot soldier has analogous meaning in this situation. It recited that the air force could bomb an enemy position with its planes, the artillery could shell it with its cannons, but until the infantryman with his rifle moved in and took control of the position, the battle was not won. The infantryman was the central figure in that case and was supported by the other groups. Everything the others did was carefully planned and executed to make his job easier. The same is true of the craftsman. He is the foot soldier in today's construction industry.

The closer to the craftsman's viewpoint the manager can get in anticipating activities and conditions, the more efficient his support plan will be. When he succeeds, units of work per man-hour of labor—*productivity*—will increase over what it would otherwise have been simply because the right things happened, or were at the right place, at the right time.

In so doing, however, the manager accomplishes something far more significant. In providing direct support, he also creates an efficient operational atmosphere, which, in itself, is a moti-

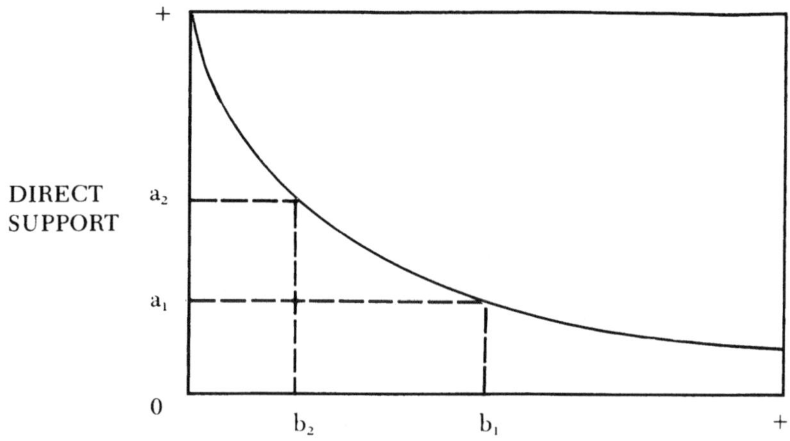

Figure 1-3 Direct Support vs. Frustration Level

vating influence on the craftsman because he knows he will not experience a lot of frustration due to lack of planning. As direct support of the craftsman *increases* from a_1 to a_2, his level of frustration *decreases* from b_1 to b_2, and as frustration falls, motivation *increases* (Fig. 1-3). Experience in the industry indicates it is a basic fact: the craftsman enjoys his craft, and the easier it is for him to exercise his skills, the more he will produce in terms of safe and defect-free units of work in a given time period.

Psychological Nourishment

In motivating themselves to produce to high standards of achievement in terms of unit cost, quality of workmanship, and accident-free performance, today's craftsmen have needs that go beyond the needs satisfied by direct support. Since these needs are more subtle, they are somewhat more difficult for the manager to satisfy, and involve the development and use of applied psychology skills.

Earning a sustaining wage is not the challenge to the current generation of craftsmen that it was to their forebears. This evolution has been brought on by a number of independent factors, including minimum wage laws, fringe benefits such as medical insurance, and global economic growth. Craftsmen today are also better educated and more aware of their surroundings. They therefore demand more psychological nourishment than did their predecessors. When they obtain it, they become stronger both intellectually and psychologically and thus able to achieve higher goals in almost everything they undertake, including their activity in the workplace.

The manager's challenge, then, is to create the proper psychological environment, one in which the craftsman wants to work, and to work productively. This involves empathetic reasoning and modifying the work environment to suit human needs. The manager must develop an understanding of craftsmen's needs: the need to belong to something—a well-run construction organization, in this case; the need for esteem—both self-esteem and the esteem of fellow workers and supervisors; and the need for the recognition that comes, or should come, with accomplishment.

Further, the manager must understand the craftsman's need to be recognized as a living, breathing individual and not just a reservoir of foot-pounds of force at the end of a hammer or shovel, who is controllable with words instead of levers and push buttons.

In other words, the manager must be able to humanize the workplace. This will become significantly more important as robotics move out of our factories and onto our construction projects in years to come. Craftsmen will demand treatment that clearly separates them, their needs, and their output from that of machines. The successful manager will be the one who can anticipate these needs and take the proper creative action. This will involve recognition by the manager of the features and characteristics that separate a man from a machine.

The similarities are readily apparent—both can dig holes and lift loads—but the dissimilarities many times escape the manager. A machine needs fuel and oil once a day, but this can be done virtually any time, any place. Craftsmen, on the other hand, need water and food frequently during the day to replace the quantities of energy worked off, and they need to partake of this "fuel" under conditions that are, at the least, sanitary and sheltered from the elements.

Today's craftsmen also need to communicate. They need to understand management's objectives; they need to know what the plan and assignment are; they need to be able to voice their suggestions and gripes to someone; they need to be told when they have done something right instead of being told only when they have done something wrong. They need acknowledgment of the fact that they are human beings, and they need to know that they will be recognized for their contribution to the team effort. In summary, the manager is faced with the task of satisfying the craftsman's psychological needs while minimizing any factors that will turn the craftsman off. The manager must, as the old song goes, "accentuate the positive, and eliminate the negative." These actions will establish the correct climate for craftsmen to motivate themselves.

JOB RATING

Good craftsmen can visually rate a job site quickly and accurately in terms of the direct support given to the workforce. They will stay on a poorly rated job site only long enough to locate other work. As soon as they can, they will leave because of the frustration they know they will experience. If no other projects are available, or if they are tied to a particular area for family or other reasons, they will stay on the job, but one of two things will happen.

1. The frustration they experience in not being able to do the work they are trained for and enjoy will lower their motivation.
2. They will observe that the standards of acceptance—in terms of workmanship, productivity, and safety—are low, and they will, consciously or not, lower their standards to conform to the project. The end result is demotivation—the "I don't give a damn" syndrome.

In each case, the result is lower productivity. In many cases, it also results in poor workmanship, increased amounts of rework, and increased costs. It can also lead to higher accident rates, with resulting injuries and fatalities.

2

DIRECT SUPPORT

The first action a manager should take to assure maximum overall productivity is to develop an operating condition in which the performance of each individual aspect of the work is efficient. A norm should be created in which all operations are perceived by the groups involved to be snag-free from the time materials arrive on the job site to completion and check out—ready to use or operate.

To do this, the manager should develop an organizational philosophy, a sense of responsibility in management, supervisory, and staff sections that causes members to want to perform their assignments effectively. Collectively, they should be dedicated to getting everything ready so the craftsmen can do their jobs on the first try safely and without delay.

This sense of dedication is sometimes difficult to develop because of the manner in which management has viewed the craftsman in the past.

MISTAKEN PERCEPTIONS

Some noncraft employees in the construction industry continue to characterize the craftsman according to McGregor's Theory X as stupid, lazy, and unconcerned about productivity. They view craftsmen as the least important link in the organizational chain, as not really mattering in the scheme of things. This negative perception of the craftsman is usually held by those who possess one or more of these characteristics themselves and draw comparisons in order to bolster their own egos. This group can be broken down into two subsets: the "negative-positives" and the "negative-negatives."

The "negative-positives" (those who are negative about the

craftsman and positive about themselves) see their own roles as ends in themselves and accordingly feel that only what they do is important. This type tends to be egotistical and usually fits an easily recognizable behavioral pattern. They genuinely believe that everything that has taken place before their particular contribution is but a preliminary to the essential activity—their contribution—and that any activity following theirs is minor and of no real significance. Unfortunately, many bright, highly trained people progress quite far into their professional careers before they recognize the fallacy in their logic.

The manager's task, in this instance, is, first, to recognize the symptoms of such an outlook and, second, to develop a means of persuading employees suffering from these symptoms that the success of any team effort depends upon each member cooperating with all other members of the team to reach the objectives of the effort.

The "negative-negatives" are down on the craftsman and also feel that their own contributions are so minor and insignificant that it makes little difference whether or not they get the job done at all, much less on time or with any degree of excellence. This employee is usually buried so deep in the organizational hierarchy that the clarity or sharpness of the organization's goals has been worn down by verbal erosion as transmittal of them tumbles down the organization. Usually this is due to lack of commitment to the goals or decreased communications skills in middle- and lower-level supervisors. The manager's challenge in this case is, of course, to help these employees see their efforts as vital and part of a larger team undertaking that makes it possible for the craftsmen to do their job.

In dealing with both "negative-positives" and "negative-negatives," the principal activity of the manager should be to bring these team members to an understanding of and a commitment to their roles as supporters of the craftsman.

THE CRAFTSMAN

Centuries ago, craftsmen had no assistance in procuring materials, obtaining tools, scheduling operations, or paying bills. They did all of these things themselves, as well as performing the actual labor. If they built houses, they may have had their sons help them, but, essentially, they performed all of the actions necessary to complete the project themselves. Only as projects became larger and craftsmen recognized that time was an important element that could be conserved by obtaining assistance did they seek others to help them. And that is how construction support functions got their start. People began helping craftsmen by performing necessary ancillary activities so the craftsmen could be free to do what they did best.

The individual craftsman has survived to the present day. A master plumber, for example, given the job of installing the piping for a home, looks at what has to be done and determines how to do it; that is, engineers it. The plumber also performs the estimating function by figuring out how much time the work will take and obtaining prices on the materials. The job is scheduled for a time when the plumber will be free and the materials will be ready for delivery. At that time, the materials are purchased, delivered to the job site, and warehoused safely until the plumber is ready to install them. When it is time to do the job, the plumber makes sure he has the correct wrenches, solder, clamps, and hangers on the truck; arranges for a backhoe if a trench must be dug; and, after all this preparation, performs the actual labor. And as the job is executed, he keeps track of his time and expenditures so he will know what his costs are versus what they should be at each point along the way. Every major construction-related function—engineering, planning, scheduling, cost control, purchasing, traffic, warehousing—he has performed himself, in addition to the labor itself.

Could other construction-project team members be as successful in such an endeavor? Probably not. Consider, for example, the purchasing agent. The purchasing agent performs a critical material-control function because nothing comes on the project site unless he orders it. He is responsible for spending more than half of the money spent on a typical construction project. But could he determine what the job requires in detail or how to do the job? No, because he has no training in technical aspects of the craft. Could he obtain prices on the materials? Yes, because that is an area in which he has training. But who is going to tell him how much and what type of materials to buy? Does his experience prepare the purchasing agent to determine how long the work will take? No, and it doesn't prepare him to determine the methods and best equipment or tools he would need, either.

The key point is that while everyone else on the project has special skills that contribute in some way to getting the job done, the craftsman, because of history reflected in both his formal and his on-the-job training, is the only one who could probably do the whole job himself. The others therefore support him and perform functions that assist him, not vice versa. They do this so he can be free to do what they cannot: perform the labor.

MAJOR CATEGORIES OF SUPPORT

Direct support of the craftsman means seeing that everything needed is brought to the physical location at which a particular operation is to be performed at the proper time and in the correct sequence so that the craftsman can do his job. Proper direct support will naturally help the work flow more smoothly, and it will also preclude many of the frustrations that can shatter the craftsman's motivation. There are four major categories of direct support of the craftsman.

1. work planning
2. supervision
3. materials
4. tools, equipment, and supplies

Work Planning

The term planning, as used in this book, means planning the what, where, and when of project execution. Planning has two basic components: (1) technical planning, which includes engineering, architecture, and similar technical disciplines, and (2) time planning.

TECHNICAL PLANNING

Most people do not think of engineering and architecture primarily as planning activities. Close examination, however, shows that they are indeed the ultimate planning functions. For example, the engineer plans the size of beams and columns so that they will support the eventual loads placed upon them; he plans the diameters of piping so that it will carry fluids adequately; he plans the location of buildings and equipment so as to coordinate properly with other work. The drawings on which the engineer lays out his ideas are even called *plans*. The basic design is, of course, the means by which the design engineer contributes his support to the craftsman. He is trained to perform the calculations, comply with the codes, specify the correct materials to do the job, and satisfy the intent of the owner of the project. Questions regarding what is to be done, how much, and where are answered in his drawings and specifications and the subsequent details that grow out of them. On well-engineered plans, this information may also include special sequences of installation the design engineer believes would be helpful to the construction effort or any special methods that are necessary.

It is important to recognize, however, that in most instances, the output of the engineer must be amplified to allow the design to be actually executed in the field. Although the engineer usually delineates the fundamental design of the work, he looks to the vendors and contractors to flesh out the design required to make fabrication and installation possible. Jobs customarily left to the vendor and contractor include the design of temporary work, such as structural falsework and concrete forms; reinforcement placing drawings; structural connections and dimensioning; piping layouts; and wiring schematics. In many instances, the craftsman can and does perform the detailing function; but in cases in which considerable time is required to complete this activity, it is usually more economical to support the craftsman by having a detailer-designer on the contractor's staff complete the activity.

Design and detailing together—collectively called *engineering*—should communicate clearly what the craftsman is to do. Drawings used by the craftsman should be so legible and completely dimensioned that he doesn't have to spend a lot of time under adverse job conditions which, on some projects, may include wind, dust, rain, noise, and heights above or below ground, doing arithmetic to find the missing dimensions. It is far less expensive and much faster for an engineer, design draftsman, or detailer working in a relatively quiet office with a personal computer or an electronic calculator on his desk to perform this support function.

It also benefits the craftsman when designers and detailers avoid abstract references to design standards that they themselves do not even have at their fingertips. If, for instance, the designer wants 3 inches clear cover on reinforcing bars in a concrete footing, it is simpler for him to show it on the drawings than to rely on some reference in the general notes that include some twenty or thirty technical-society design standards and specifications.

From another perspective, the design engineer has to protect himself with regard to errors and omissions, especially in today's litigious society. Reference to such standards will tend to ensure this, but the little time it takes to put job-specific information on the drawings or in the specifications will save ten times that amount in the field.

Those performing the engineering should recognize that their products—drawings, specifications, and details—are not ends in themselves. The project is not finished when the engineer puts the drawings in a mailing tube for transmittal to the field; actually, in terms of funds expenditure and physical manifestation, the project is just beginning.

There are two important points to emphasize here. First, the engineer's contribution, although vital, is only part of the total effort. On most projects, it represents less than 10 percent of the total cost. Second, the engineer is the first link in a series of events that will eventually result in a completed project. Very little can take place until the engineer performs his role. Both points should be kept in mind when one considers again the overall message: the engineer is there primarily to assist the craftsman by performing design services, and he must provide his support in a timely manner.

Time Planning

Time plans develop answers to when and in what sequence a project will be executed. This information is commonly portrayed in logic diagrams and schedules. The extent of coordination between the entities—principally engineering and construction—involved at one time or another on a project varies with how the owner prefers to or must do business. The design engineer may or may not be involved during the construction stages of the project, depending on how the project is structured. For example, most government and public works projects are, by law and contracting procedure, implemented through the award of a lump-sum (a fixed price to perform all the work included in the scope of the project) contract based on competitive bidding on completely engineered drawings and specifications. Obviously, in situations such as these, there is little opportunity for the engineer and the contractor to dovetail their operations to reduce project-time parameters and resulting cost.

In industrial and commercial applications, however, it is common for an owner to bring the contractor in as soon as 20 or 30 percent of the basic engineering is completed so that the considerable benefits of cooperation and overall cost reduction, in addition to time conservation, can be achieved. In many cases, a year or more can be eliminated from the overall timetable for a project when dovetailing, as explained below, is used. This means that the project owner can begin selling the product the plant will manufacture a year earlier, thus shortening the payback period for the funds committed. In addition, because of the competitive aspects of American industry and the rapid advance of technology, a year's lead in marketing a new product can spell the difference between success and failure insofar as market share is concerned. The same principle applies to commercial construction. The first owner finished with a new office building or shopping center in an area will usually be more successful than will those who follow, other factors being the same. It is this more prevalent commercial-industrial case that will be focused on here. Also, even in the case in which the engineer is no longer involved or is only peripherally involved through drawing approvals and similar activities, actions of the construction contractor with regard to planning should be directed toward the same logic and time objectives described in this book.

SCHEDULE

When a project is conceived, its owner's first two questions are (1) how much? and (2) how long? The first question is answered by the preparation of a cost estimate, and the second question is answered when an overall design-procure-construct schedule is developed. Unless the engineering is to be entirely completed before construction begins, most efficient conservation of time is achieved when the engineer and the construction contractor first produce their own schedules and then merge them together in an optimized overall project schedule—one to which they both commit.

A project schedule can be created from a million different combinations of elements, one of them as viable as another, but the schedule to which the engineering and construction entities commit themselves is the one they agree can be held to. This is the starting point from which all subsequent activity ensues. Commitment to a common plan means that the engineering and construction entities will act responsibly to perform their parts as agreed. If each does its part when the schedule calls for a particular activity to be performed or a specific decision to be made, that part of the work will be completed on time; and if all the activities are performed on time, then the project will be completed on time. Sometimes, through lack of communication, this simple but important concept is not acknowledged or even recognized.

The common planning tools for a project are the several levels of schedules that are developed. At the highest level is the master control schedule. It is the document that serves as the wedding ring in the marriage between engineering and construction. It is the document that confirms the agreements reached and the commitments made by each group so that a project can flow smoothly from the design boards to the field.

Commitment to a common schedule is the only way construction on a project can start before the drawings and specifications are 100 percent completed. This approach saves valuable time, however, because of the tight intermeshing of engineering, procurement, and construction operations it achieves.

Intermediate level schedules for both engineering and construction are developed from the master control diagram. They go into successively more detail and cover smaller and smaller intervals of time. On the construction side, top-level project management should develop the overall plan and the highest level of schedule. Each successively more detailed level of schedule should then be prepared by, or with heavy participation by, the corresponding lower level of supervision.

The most detailed level should involve the foreman, and the foreman, in turn, should involve the craftsmen. There are two reasons for this. First, the foreman is closer to the work and knows more about the details of the operation and the capabilities and limits of himself and his crew than anyone else. He is thus in the best position to plan the work, coordinate with others, and make the decisions about what tools, equipment, and supplies are needed. Second, his involvement and the involvement of the craftsmen brings commitment at the grass roots. It induces the craftsmen and the foreman to keep their word; to do what they said they would do. Nothing gives a good craftsman more self-satisfaction than achieving a particular goal by the time he said he would. Involvement in the planning process brings commitment, and commitment brings an increase in self-motivation. The effect is self-propagating.

This deepest level of detail is usually a simple bar chart that the foreman carries to serve as a guide. It details every operation the foreman and the crew will perform in the current week, and it shows what's planned for the following week. It is the document the foreman uses to allocate manpower, to requisition tools and supplies, and to schedule equipment needs. It is the way he assures that the things under his control will be done so that the craftsmen can do their job without being delayed by missing items.

On a large paper-products project, the owner decided to maximize the efficiency of the dovetailing benefit by bringing the contractor into the picture when the engineering was only 20 percent complete. Faced with a limited budget and a schedule dictated by a skyrocketing demand for the product to be produced by the completed facility, the owner participated directly in the detailed planning of the project. He directed that each component for which a specific design was required be identified and listed separately so that it could be coordinated and scheduled closely—*dovetailed*—for installation in the field. The construction contractor had planned his schedule carefully so that he could progress the work on an area-by-area basis. He wanted to complete his work and move the craftsmen, tool bins, equipment, and storage on to the next location without having to return to finish little bits and pieces in a time-consuming and costly manner. The engineer reviewed the requirements thoroughly, compared them with his capabilities, and met with the contractor to eliminate mismatches and conflicts. Upon completion of this coordinating activity, the engineer and the contractor committed themselves to the common schedule. The engineer encountered some difficulty in gaining genuine commitment from his staff, however.

For example, the schedule called for the boiler exhaust stack foundation to be designed by the end of what was called Week Number 43. (The schedule used a project week numbering system to avoid confusion and missed commitments. In it, each week was assigned a sequential number rather than a calendar date. This was because one firm used Sundays to designate the beginning of calendar weeks, another used Mondays, and a third worked with end-of-week dates, using Fridays. However, they agreed that there was only one Week Number 43, the forty-third week after the date when the project was first initiated.)

During one project review meeting, the designer who had been assigned responsibility for the stack told his supervisor that he had fallen behind and was not going to be able to make the schedule on this relatively minor item. It was not a big problem, he felt, and he would complete it when time permitted. The supervisor reminded him that with an overall schedule as tightly coordinated as this one, it was imperative that the engineer finish the design on time. In response to the supervisor, the engineer said he couldn't accept the argument that the project was going to suffer because some small item like a boiler exhaust stack foundation did not get designed on time; he would get it done sometime before his work on the project was completed.

His supervisor patiently replied that this was the schedule the designer had drawn up for his own activities, one that had been integrated into the rest of the project schedule and one to which the construction contractor was committed. Further, if the designer delayed completion of the design to the following week, some activity he had set out for the following week would have to be put off until the week after that, and so on. The end result would be that the overall project would be a week behind schedule.

There were, no doubt, ways in which a real slip in schedule could have been avoided, if necessary. The designer could have been asked to work overtime, or someone else could have been assigned to do the work. But the point the supervisor was making was important from an attitudinal point of view. Although there were many different combinations of activities that might have gotten the job done, the schedule to which everyone was working was the one that all involved in the project had agreed on. The contractor's field operations—excavation, form building, reinforcing fabrication, anchor bolt procurement, and the like—were dependent on the design being completed during Week 43. And in the field, this particular foundation was in such close proximity to several others that it had to be constructed in a given sequence, one that had been developed as part of a master concrete-placement plan during the original scheduling process. To skip over it would have required bringing all the involved craftsmen—laborers, carpenters, rod busters, and finishers—back to that point for one foundation, thus delaying back-fill operations and the stack erection itself.

After thinking it through and looking at the requirement from this broader perspective, the designer decided that he would be able to finish the design on time after all. Once he realized that many others were depending on his performance and that his efforts were vital and an integral part in supporting the craftsmen in the field, his commitment to meeting his part of the schedule became resolute. He realized that his role in the project activity was important and vital, but not an end in itself.

The message was conveyed to the rest of the design group as well, and a satisfying team effort resulted; the project did, indeed, finish on schedule, and both the engineer and the construction contractor were awarded much more work by that client.

Each item of activity on a schedule can have an impact on

the outcome of a project. The chain of events for any item is substantially the same. The schedule, after it is developed and committed to by all entities, should be expanded in detail and distributed to every discipline on the job so that not only the line departments but also the staff functions—purchasing, warehousing, safety, and others—will be aware of what is happening next week and next month. The benefits are obvious. The safety engineer, properly informed ahead of time, can prevent unsafe acts and conditions from occurring when the work begins. The procurement and warehousing departments can assure that adequate quantities of materials, such as form lumber, pipe, and fittings, are on hand and also make sure that deliveries of fabricated materials and equipment, such as structural steel, pumps, and transformers, are expedited to support the work in the field.

On a typical industrial project, the construction phase is comprised of excavation; the placement of foundation forms, reinforcing steel, concrete, and structural steel; process equipment installation; and piping, insulation, electrical, and instrumentation work. In most instances, these operations must follow each other in a precise sequence; it is therefore critical that the schedule of delivery of materials and equipment to the job be followed. If schedules begin to slip, corrective or adaptive action must be taken based on accurate, up-to-date information.

Once a plan is put into operation, it must be updated frequently to reflect any delays. Failure to monitor equipment and material delivery status closely can permit slips in deliveries that cause losses far more significant than simply loss of time. Just as important as monitoring and updating is careful communication of any changes in the plan to everyone involved in the project, including those in support functions, first-line supervisors, and the craftsmen themselves.

As a rule of thumb, even the entry-level craftsman apprentice should be aware of, and be able to plan for, the things he will be asked to do on a particular day. The journeyman he assists should be aware of what is going to be done the following day as well so that he can plan out his activity before he gets to the job the next morning. The crew foreman should be at least a week ahead with his task planning, and he should have a good mental picture of the activities planned two weeks ahead. The general foreman must determine his manpower needs and his requirements for tools, equipment, and supplies at least a month ahead of time so that these resources can be obtained; and the superintendent must be able to see the whole project as it relates to his craft. The support functions—personnel recruiting and hiring, timekeeping, safety, and warehousing—should also be made cognizant of any changes in the schedule, especially if it means a change in craft manpower needs. Running out of a needed material can shut a project down faster than can any other single factor.

A herbicide plant project once started to fall behind schedule because deliveries of critical process equipment were slipping. The manufacturing schedule had not been closely watched because, in this instance, the customer himself was manufacturing some of the equipment in a factory overseas and had given the contractor assurances that delivery would not be a problem. Unfortunately, however, although the customer's project representatives had committed themselves to the overall project schedule, they were unable to obtain the same level of response from their manufacturing counterparts.

It was imperative that the completion schedule be achieved because the product the plant was to manufacture was seasonal. Approval by the company's top management had been contingent on the project being completed as planned. Careers hung in the balance. Virtually an entire year's production of herbicide had to be manufactured and stored for shipment by the spring of the year. The economics of the venture depended on having the product for the scheduled season, not a year later.

The project manager brought up the subject of the slipping schedule in a meeting with his construction superintendents. He told them that the only way to overcome the delay resulting from late equipment deliveries was to increase the size of the work force by adding more welders and pipe fitters. He told them to hire the people.

Approximately fifty additional craftsmen were recruited. A problem arose, however, because no one had thought to tell the support people in the supplies warehouse of the increase in manpower and so there were not enough welder's helmets and safety glasses to go around. By the time this was discovered, the men had already been hired and, as a result, spent their entire first day on the payroll sitting in the shade waiting for the necessary safety equipment. Certainly the money paid to these men for sitting around represented a loss, but a more expensive cost to the project resulted from the psychological effect the oversight had on the men. They had been recruited to go to work—to fit up pipe and weld—and as a result of management errors, they could not do their jobs. Even after they had started work, they retained reservations about the job. Little problems became magnified in their minds because of their initial experience. Frustration flourished. Motivation decreased. Formation of a positive attitude about the job had been precluded before the men even started working.

The original development of a plan is the easier part of planning, and many managers let it go at that. Revising the plan to meet or anticipate the changes that invariably take place in any project is more difficult and time-consuming. Finding out what is happening or is about to happen as a result of the many influences that come into play as a project proceeds is a continuing challenge. Many supervisors do not realize how vital to efficient job performance updating the schedule can be.

When a project falls behind schedule, even the best "catch-up" (remedial) plan will fail unless it is properly communicated. The effectiveness with which the manager and the staff communicate any changes in the plan to each other and to the rest of the project organization will determine the degree of success the project eventually enjoys.

Keeping the plan up to date on a construction project is similar to the way a football game is planned and executed. Before the game, the coaches and players formulate a plan—the *game plan*—to guide them in their preparation. This plan reflects their best collective knowledge of the obstacles that will face them on the day of the game and how to overcome them. This is analogous to the overall schedule prepared for a construction project. It represents how those involved agree that the project can be carried out most effectively.

During the football game, the players huddle or meet before each play to decide which play will be run to gain the most yardage. As the players go to their positions, they know their assignments. On a construction project, a planning meeting, similar to a huddle, is held, usually each week. During this planning meeting, the players—managers, superintendents, and staff personnel—review the overall plan for amount accomplished in the previous period, and determine ways to maintain or improve the performance. Each player leaves the meeting with a clear understanding of his responsibilities during the coming week and how these responsibilities mesh with the activities of others on the project.

In a football game, when the actions and positions of the opposing team are changed or are not, in some way, what was expected, the quarterback changes the play at the line of scrimmage by calling out an "audible" signal. Hearing it, each player revises his own assignment in accordance with the new play. Similarly, on a construction project, when some influence—weather, for example—requires a change in plan, the people affected should be told immediately of the change so that they can adjust their assignments to minimize lost time and effort.

Unfortunately, there are times when the noise from the crowd in a football stadium causes some of the players to misunderstand or miss entirely the "audible" signal. When this happens, players go in wrong directions, assignments are missed, and the play is a failure. Crowd noise, over which the quarterback has no control, can thus cause a significant breakdown in the communication of a plan or a change in plan. A construction project is better off in this regard. Construction projects today have virtually unlimited instant-communication capacity via the telephone, a radio, or a computer. There is little excuse for mix-ups ("busted plays"), which must be blamed on a failure by management to assure communication of planning changes. There is no "crowd noise" to blame it on.

A well-managed project is one that is viewed by its project manager from a dynamic prespective. He recognizes that the needs of a project are in a continuous state of change. These changes are brought about by both external and internal influences. As in a football game, strengths and weaknesses may be discovered in the opposition as well as in the team itself. The weather may change; it may rain or snow. A critical piece of equipment may be damaged in shipment; a key manager may get sick or leave for a better job.

All these factors tend to delay the progress of the project toward its goals. An ordinary manager will shrug his shoulders and conclude that these are things over which he has no control. He will eventually fail. The innovative manager, on the other hand, knows that the main challenges he will face are the unexpected, and so he will be ready and waiting for them when they come. He understands that his primary responsibility, like the quarterback's, is to recognize the dynamics of the situation, what the threats are, and where advantages may be gained; and he will implement adaptive or corrective action as required to keep the project on a steady, safe course toward its goals of on-time completion, under-budget cost, and workmanship to specification.

Completing a project from ground breaking to commercial use is like driving a car on a long, straight stretch of highway. At the beginning, it looks simple—nothing but an easy road ahead to the far end. However, farther along, a bump on the road may cause the car to lurch slightly to the side, requiring a slight correction back toward center. Later, something else—perhaps a strong gust of wind—may cause the car to veer off center again, requiring another slight correction back toward center. Even though the end of the road is clearly in sight, from the beginning outside forces from many different directions may cause the car's progress to be not a straight line, but instead a series of corrections which, if not taken, would turn the trip off course and toward disaster.

In the case of a construction project, the external forces may include changes in design, which cause additional work; *holds*—orders from the engineer to stop work on a portion of the project; late shipment of equipment and materials due to tardy procurement or manufacturing slippage; and incorrect or incomplete shipments. In some instances, the quantity or quality of labor may be a problem. Sometimes even a shortage of support personnel—supervisors, buyers, or safety engineers, for example—may be severe enough to slow a project down. The astute manager is aware of these potential problems and is ready to implement adaptive or corrective action.

One manager, building a fertilizer plant during an especially busy national economic cycle, saw that because of the rush of orders, the manufacture of his equipment was slipping behind schedule. He sent an engineer to a plant where much of his equipment was being made to assess the situation and determine if authorization to work overtime should be given. The engineer reported that the plant was extremely busy and was already operating on an overtime schedule. He reported, too, that other customers, also experiencing delays, had assigned full-time expediters to the plant and that the pressure of their presence was yielding results with the manufacturer.

Realizing that the "squeaking wheel was getting the grease," the manager responded to the severity of the problem by reassigning half of the project's engineers to be expediters. He also made the engineering manager the expediting manager. The new expediters were dispatched to each factory in which manufacturing or fabrication orders had been placed. Some of them became full-time residents at a single plant location; others visited their assignments several times a week to verify the status of their orders. They identified steel, castings, pressure parts, valves, and other equipment, marked them, copied down serial numbers, and otherwise prevented those items from being appropriated for use in other customers' orders. They obtained copies of production schedules and raw-materials purchase orders to verify that procurement and day-by-day rolling, casting, machining, and welding operations were performed as scheduled. Slippages were reported immediately, and upper-management pressure was applied as required. Phone calls between company presidents regarding details normally handled at clerical levels became common. Design engineers were dispatched to the manufacturer's location to make required drawing reviews and approvals so as not to lose transmittal time in the mails. Several cross-country rail shipments were escorted by the assigned engineer-cum-expediter himself to prevent delays along the way. He actually rode in the caboose with the train crew. One shipment so escorted traveled from Seattle to central Florida, location of the project, in twelve days instead of the six weeks the trip normally consumed. The end result was that the equipment-delivery slippage was recovered and the project was eventually finished on schedule.

Internal delaying or disruptive influences also require constant recognition and action. Consider, for example, the location and staffing of supply rooms. Supply rooms and tool cribs should follow the work force in location as well as in size and staffing. At the beginning of a job, only one central toolroom may be required; at the peak, three or four may be required. As the activity of a project progresses from site work, underground, and foundations to building finishes, equipment installation, and piping, location of the tool and supply rooms should be reviewed to determine if they should be moved closer to the centroid of labor, the place where most of the action requiring travel to them is located.

In addition, if a craftsman must stand in line more than a minute or two to withdraw a tool or pick up supplies, then additional toolroom attendants should probably be assigned. If he must walk more than three or four minutes to reach the tool crib, then additional tool cribs may be required, or the existing ones may need to be located closer to the work area. The same holds for portable toilets, lunch areas, and so on. Every minute a craftsman spends walking to or from his workstation is a minute during which he is unable to do the job for which he is trained and for which he is being paid. On a poorly managed project, traveling costs can amount to as much as 30 percent of the cost of labor.

The alert manager can improve productivity on his project simply by observing the sources of delay and taking the actions necessary to make the work flow more smoothly. The bonus, however—and this frequently even outweighs the impact of the direct-support actions—is that a smoother work flow reduces frustration in the work force. This has an intangible but real impact on self-motivation.

A good craftsman, the type who is vital to the success of a project, has many personal qualities in common with the good manager. He is personally well-organized. He likes to see things run as anticipated. He takes pride in his work. He gets immense ego satisfaction out of planning an activity, performing it, and having it turn out as he intended. He is trained, experienced, and ready to deal with many of the factors that tend to prevent him from doing his job efficiently. To him, that is part of being a craftsman. However, he becomes frustrated and discouraged when factors over which he has no control affect his progress, especially if he believes that these factors could have been prevented. Understanding of this concept, critical to the success of any motivation program, is discussed in more detail in Chapter 7.

Supervision

One of the most important elements necessary to the support of the craftsman is the direction he is given by his supervisors. The supervisor acts as a leader, a source of craft knowledge, and an advocate.

As a leader, he directs the craftsman in performing the work. He coordinates the other essential elements involved. He is the person responsible for assuring that the plan, the materials, the tools, and the labor are conveyed to the point of the work at the right time and in the sequence proper for that particular craftsman or crew of craftsmen. His job is to meld these diverse elements together into a package of energy that when released, will accomplish a portion of the work at hand in the most efficient manner possible. This is true of all supervisors, from the foreman up to the job superintendent or project manager.

The foreman, who is the first level of supervision, must possess the basic supervisory skills: he must know the principles of his craft well enough to command the respect of his subordinates and to maintain the standards of workmanship set out by the specifications. He must be able to lay the work out, show the craftsmen how to do it, if necessary, and give accurate answers to their questions. He must know what tools and equipment are required to do the work and must be familiar enough with project operational procedures to obtain them when needed. He gives the craftsmen information on correct safety procedures and instruction on more cost-effective ways of doing the work. His role is to draw upon his experience in the craft to assist the craftsmen in performing their roles, the installation of the work. He must empathize and understand what motivates or demotivates craftsmen.

It is also at this first line of supervision that management is represented to the craftsman. If the foreman is knowledgeable, fair, and organized, management will be perceived similarly. If the foreman is weak, two-faced, and inattentive to detail, the craftsman's perception of management will be the same.

At the next higher level, general foreman, the supervisor must have an understanding of what is to be done over a wider area or over a number of small project components. He usually must be capable of helping three or more foremen coordinate the efforts of their crews. He must have a broader knowledge of construction methods and must guide his foremen in selecting the best methods for accomplishing given tasks.

Still higher, at the craft-superintendent level, the supervisor must take responsibility for all work in a given craft on the project. This means that he must maintain the cost, schedule, quality, and safety objectives as they relate to his craft. His support of the craftsman is somewhat more indirect, but nevertheless vital, because he provides overall direction and coordination to all craftsmen in his craft. He makes sure there are enough craftsmen on the job to do the work safely and to maintain the schedule but not so many as to make it inefficient. He also guides the general foremen in determining the crew size required for the particular operation at hand and the mix of skills needed. In planning the execution of the work, the craft superintendent seeks to answer questions such as: Can the job be done with two craftsmen, or does it require eight? Do all of them need to possess journeyman-level skills, or can a portion of them be helpers or apprentices? Is there a special requirement for coordination with other crafts or other crews in the same area?

When a larger crew is needed to perform some task, it is the craft superintendent who determines whether to transfer men from another crew or add craftsmen to the payroll. He looks at the overall picture in his craft and plans manpower levels to support the project schedule. He sets the standard of acceptance regarding craftsman skills. He also determines what basic installation methods and techniques will be used and what tools, equipment, and supplies will be needed. He then makes sure that these items are available at the point of the work when the craftsmen need them.

At the top of the job organizational hierarchy is the job superintendent or project manager, who must coordinate the whole job. It is sometimes difficult for craftsmen to see how this level of supervision—the "big boss"—helps them at all. This is because the foreman usually spends virtually all of his time supervising his crew and almost none of it managing all the other elements of the work, as illustrated in Figure 2-1, points a_1, b_1, and c_1, while the project manager or job superintendent does just the opposite. He spends virtually all of his time managing other elements of the work, both directly and through his subordinates, and almost none of it supervising a crew as represented by a_2, b_2, and c_2. He is simply not as visible. His efforts go largely unnoticed by the craftsmen because they involve working through other people.

The supervisor at any level also plays another important role, that of advocate for the people who report to him, whether they be other supervisors or craftsmen. In this role, he is a communicator. He listens to suggestions, complaints, and gripes and

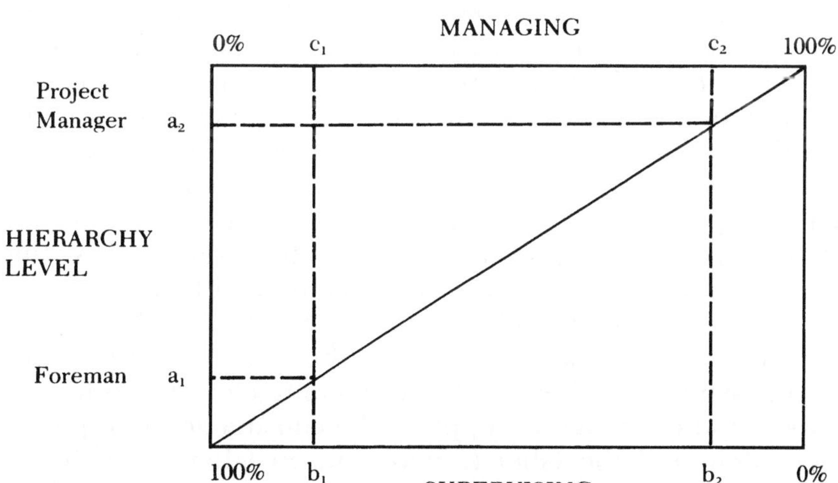

Figure 2-1 Management Level vs. Supervising Crew Activity

sparks action when indicated. He also evaluates the craft and supervisory skill levels and the attitudes toward working and safety. To play this role, he must know his subordinates and their job responsibilities well enough to empathize with them and to see problems from their viewpoint.

Materials

Construction is installation of materials and equipment. The craftsman can install only what he has been furnished. If he has everything that is required, he can complete the installation. If materials or parts are missing, he has to wait. Waiting at today's labor rates gets expensive very quickly. In rare instances, he can fabricate a missing piece or part—assuming the specifications will permit it, he has an adequate stock of raw material on hand, and he has the tools with which to do it—but this is also usually more expensive.

Attention to detail in the procurement of materials, supplies, and equipment is therefore mandatory if labor is to be properly supported. Delivering only a portion, or even most, of the materials required for a particular task is not good enough. Everything required to complete an installation should be there at the time it is needed and in the proper sequence. What happens when the correct quantity, size, and specification of pipe is delivered to the location from which it can be most efficiently installed but the gaskets are missing? Or the connection bolts are there but are the wrong length? The crew or crews remain idle until the forgotten or correct materials are obtained or some substitute work for the day is hurriedly assigned.

On one petrochemical plant project, the plans called for the installation of an underground fire-protection system consisting of cast-iron pressure pipe. The design engineer had been commissioned by the owner of the project to purchase the process equipment and commodity materials (those normally purchased in bulk quantities). The construction contractor was to furnish the tools and construction supplies, the construction equipment, supervision, and the labor to make the installation. This is a relatively common division of responsibilities, yet something went awry.

The engineer purchased the materials and scheduled them for delivery to the job site. The contractor planned to begin installation at once and so he had opened up trenches to prepare the bedding for the pipe. Unfortunately, it was not discovered until the pipe started arriving that the engineer had not bought the lead and oakum needed to make up the joints between sections of pipe.

The root cause of the problem, it was discovered later, was that the engineer thought that joint materials were construction supplies and thus the contractor's responsibility. The contractor, on the other hand, assumed that all the materials required to make the installation would be furnished by the engineer. Lack of a clear understanding between the engineer and the contractor with regard to the scope of supply had created a situation at the construction site that forced a stoppage in the work until the missing materials could be obtained. Because that type of joint was no longer in widespread use, it took almost two weeks to get quantities of lead and oakum sufficient to continue the work. In the meantime, several plumbers who had been recruited away from other employment especially for their skill and experience in making up that type of joint had to be temporarily laid off. It does not require much imagination to understand their lack of motivation when they finally did get started.

Motivational Impact on Craftsmen

When the correct types and quantities of material fail to reach the point of the work in coordination with the labor, tools, and equipment, good craftsmen shake their heads in frustration; they came to work that day with an idea of what they were going to do. Each of them, consciously or unconsciously, walked through the gate with a goal for the day: to do something which, at quitting time, they could look at and say, "I made my goal today" or "I planned to make eight perfect welds and I did it." Instead, they wasted their time—and their employer's money—while alternate work had to be found for them, simply because someone else failed to do his job. Usually, craftsmen target someone in management as the culprit, and they are usually correct. Good craftsmen will not stay on a poorly run job any longer than they have to because they get no satisfaction from it—only frustration.

Sequence of Delivery

Many times, the sequence in which a project is, or must be, designed is exactly opposite to the way it must be constructed. A building, for instance, must be engineered structurally from the roof to the foundations, while its construction obviously must progress in the reverse order. In a process plant, a piping header or main line is designed after the sizes and flow rates of the smaller lines that attach to it are known, but the most efficient construction sequence usually begins with the headers first and the small branches or drops to the process equipment last.

Occasionally, due to inferior control procedures and pressure to maintain schedule pace, a sequence of ill-defined goals develops that ultimately penalizes the project by frustrating and demotivating the work force.

Consider the agricultural chemicals project that had, so top management believed, a closely coordinated schedule between the engineer and the construction contractor. The timing on the project was tight because of advance sales for the product the plant would manufacture and projected market demands. Dovetailing of the design-procure-construct sequence for each discipline was carefully fine tuned to assure minimum interruption between design office and field. The owner's contract with the engineer called for delivery of certain portions of the design by certain dates. In turn, the master schedule called for the prefabricated pipe materials to be delivered in blocks defined by percentage of quantity: 25 percent by a certain date, 50 percent by another date, and so on.

Unfortunately, in this case, the above goal definition wasn't good enough because the engineer kept his end of the bargain by delivering the first 50 percent he could (*branch piping*, the portions that attach to the process equipment) first and the headers or main lines last. This was the sequence in which he designed the piping systems but not the sequence in which the contractor had to install them.

The branch piping was scheduled to be installed after the equipment to which it attached was in place. The contractor's schedule was built around putting up the headers in each area first, before even the process equipment was on the job site. This was necessary because the header piping was of large diameter. It was heavy and bulky and had to be installed with mobile hoisting equipment that required the floor area to be clear of obstructions, such as the process equipment. This procedure also allowed piping activity to progress before, during, and after the time equipment setting was taking place. The contractor's schedule called for headers first, branches next, and tie-ins to the equipment last. Piping erection, in this instance, would be much more efficient with this technique because before each category was started, the end or connection points would already be installed. Fit-up would also be much better.

The contractor planned each part of the piping operation in great detail. Special packages of installation information, including sketches and bills of material, were prepared for each spool or section of pipe. Additional pipe fitters and welders were recruited and promised work by a certain date. Many moved to the area and rented apartments or brought trailers in so they could be ready to start at full speed ("hit the ground running," as it was called). The plan was to be ready to install the pipe directly off the beds of the delivery trucks when they arrived on the job site. Then it was discovered that the first shipments were to be branch piping—the last portion they could install.

Because of the mix-up, hiring of the pipe fitters and welders had to be delayed because there was not enough work for all of them. Several weeks later, when there were sufficient quantities of the correct materials on the job site to sustain the installation effort, the craftsmen were asked to report for work. Unfortunately for the contractor and the project itself, many of the best people had already found other work.

Besides losing the opportunity to employ some top-notch craftsmen on the project, the contractor also lost momentum that was not regained. The craftsmen who did go to work on the project correctly concluded that management was not effectively controlling the flow of the materials to the point of the work. And many of the better people left rather than face the frustrations and demotivation they knew future miscoordination might bring. They did not care what caused the foul-up or who was to blame. All they knew was that the job was not running as it should have, and they did not want any part of it.

To avoid disconnects such as the one above, agreements among the owner, the engineer, and the contractor regarding material-procurement responsibility should be carefully spelled out down to the nuts, bolts, and washers.

Even worse than shortages caused by others are those caused by contractor's organization itself. Responsibility for the accuracy of takeoffs should be entrusted to experienced construction engineers or technicians who know what is required in a given type of installation. The person making up material requisitions for concrete reinforcing bars, for example, should know enough to provide tie wire and support chairs if required. The technician requisitioning pipe flanges should make sure he has the correct material specification for the pressures and temperatures involved as well as the proper bolt length; and he should know enough to allow for replacement of pieces that are inadvertently dropped or misplaced.

Tools and Construction Equipment

Lack of proper tools and equipment at the point of the work has the same result as lack of materials: the craftsman cannot do his job. Not only must the correct tools and equipment be chosen for the work at hand and properly maintained from a safety standpoint, but equipment, especially, must be carefully coordinated between crews. In uncoordinated situations, one crew will finish using a piece while the next crew that needs it is not aware that it is available. Unfortunately, this seems to happen with the more expensive pieces, such as cranes, cherry pickers and forklifts, as well as with end loaders and backhoes.

To the craftsman, the way in which tools and equipment are managed reflects the quality of the management, and since the condition, quantity, and suitability of tools and equipment are easy to see, the good craftsman will quickly rate the job on the basis of how they look. He will make his decision on whether or not to stay accordingly. The reason is simple: the craftsman is on the job (and, in fact, in the construction industry) because he derives ego satisfaction from the accomplishment of a task that requires the skills he has mastered. He wants this satisfaction every day; it is what keeps him going. When he is prevented from doing his job by the failures of others to do their job—provide the tools and equipment he needs—he becomes frustrated. As discussed earlier, frustration means demotivation. Just as increased motivation stimulates productivity, demotivation diminishes productivity, but on a much broader scale.

Sources of Delay

Consider the dilemma of the first-class pipe welder on a poorly run job. He gets his satisfaction from the placement of defect-free beads of weld metal. That is what he has been trained to do, and he knows how to do it well. When he is prevented from exercising his skill because, for example, the scaffold that the carpenters were to build for him to climb on has not been completed or has not been completed on time, or because he cannot climb on the scaffold because the safety engineer has not put his approved tag on it yet, he becomes frustrated. He becomes demotivated. He produces less. When he finally does get to climb on the scaffold to make his weld, he stretches the job out, hoping that when he finishes one weld, the site for the next weld will be ready.

One can expect a significant decrease in the quality of this craftsman's workmanship because he no longer cares. Also, he will leave as soon as he can find other work. If this is a continuing project, the immediate result of his departure will be the necessity of recruiting and training (at the very least, on the job site) a replacement.

Projects with high turnover rates are those with poor productivity. In many cases, the reason for the high turnover rate is the failure of the job-site staff to communicate amongst themselves.

Empathize with the crew of ironworkers that has been hired to erect and weld up some structural steel for a manufacturing-plant renovation project. The ironworkers counted up their requirements and determined that they needed seven gasoline welding machines to do the job expeditiously. Imagine their frustration and disappointment when the machines, which had been leased by the manager (to save money, he thought) from a cut-rate rental equipment supplier, arrived on the job site and the ironworkers discovered that six of them wouldn't turn over because of dead batteries or broken wiring and the seventh wouldn't start because its choke was rusted shut. This was a classic case of mismanagement.

Or think about the pipe crew that had been planning to lift up a series of pipe spools and weld them in place on a particular day. When the crew was delayed because the crane to lift the pipe in place was not available, not only was time lost but the crew produced less because of the frustration caused by the delay.

Everyone is tolerant of occasional mix-ups, but when they happen all the time, craftsmen become unforgiving. Frustration-induced demotivation seems to propagate itself and compound its intensity with each mix-up.

In the above example, the pipe crew was delayed because the hoisting equipment was unavailable. When the hoisting equipment finally arrived, the crew was pressured to regain the lost time by working more rapidly. Such pressure usually results in unsafe acts or poor workmanship. In any case, being pressured and criticized by supervision and management for not producing under these circumstances is, of course, grossly unjust. The crew knows that they can place pipe efficiently, and that's what they will do if not prevented by outside influences. They may complain to management that the reason their production is down is because they did not have hoisting equipment with which to lift their materials. However, this complaint may fall on deaf ears because the manager himself was responsible—directly or indirectly through his subordinates—for seeing that the hoisting equipment was there. He may be unwilling to accept the blame personally. The risk of being criticized by *his* management is too high.

Owner's Representatives

Construction of a large refinery project was once almost strangled by management shortcomings before it even got off the ground. It was being constructed under a contract in which the owner's management had opted to proceed on a reimbursable (a method of contracting in which the contractor's costs are paid by the owner, or reimbursed, as they are incurred) rather than a lump-sum basis. In this case, however, the management misdeed was caused not by the contractor's management but by one of the owner's project-based representatives. He stormed into the contractor's office one day and demanded that a certain foreman be terminated for leaving his crew unsupervised. He had observed that the crew was without its foreman and jumped to the conclusion that the foreman had simply left the crew to fend for itself.

The project manager investigated the incident and found that the foreman had, indeed, left his crew. The reason was that they were close to running out of work because they had progressed in their assignment faster than they had anticipated, and they needed more open trench in order to proceed. The foreman had gone to find the equipment supervisor to request the equipment to dig the trench. The work site was geographically spread out over several hundred acres, and the foreman had to search out the equipment supervisor on foot and make his request in person.

This was because the same owner's representative had previously turned down several requests by the contractor for a two-way radio system that would allow the foreman to stay in constant communication not only with the equipment supervisor but also with the safety department and other necessary contacts. The contractor eventually got the tools he needed to do the job properly, but not until the owner's top management had become aware of the shortcomings in its own staff and had placed knowledgeable representatives on the project site.

MANAGER'S RESPONSIBILITY

The manager's job, then, is to bring all four of the major categories of direct support of the craftsman—work planning; supervision; materials; and tools, equipment, and supplies; and the materials to be installed—together in the correct sequence and at the proper time. The task requires imagination, attention to detail, empathetic reasoning, and decisiveness. When he succeeds in doing this, the work will flow smoothly, progress will be faster, and productivity will be higher. The most significant benefit, however, will be the increased motivation of the craftsman.

WHEN TO DETERMINE NEEDS

The time to start determining tool and equipment needs is when the labor needs are projected—during the initial planning stages of the project. After the overall schedule is developed, the project's managers, craft superintendents, and schedulers should plan equipment and tool needs as a parallel activity with forecasting labor needs by craft and skill level. The initial cost estimate sometimes provides a starting point for both of these activities, but unless the people responsible for building the project are also involved in preparing the estimate, the projections in the estimate can be used only as a rough guide. They will probably not be precise enough to serve as an accurate projection of the type and quantity of either tool and equipment needs or labor needs. For example, an estimate may project pipe-fitter and welder man-hours but be of no use in determining manpower levels of journeyman versus helper pipe fitters or rod welders versus TIG (tungsten electrode-inert gas) and MIG (metallic electrode-inert gas) welders. Similarly, the estimate probably will not distinguish between gas welding machines and DC (direct current) or single units versus eight-packs (eight DC units assembled in one large module). This must be done by the supervisors who will implement the plan.

CONCEPTUAL PLANNING

With a detailed time schedule of labor manpower levels by craft in hand, however, the supervisors should be able to conceptualize each major operation and determine the basic methods of construction. Will the foundations be mass excavated by scrapers and dozers or dug individually with end loaders and backhoes? Will concrete be placed with pumps, conveyers, crane and bucket, or will it be chuted? Will structural steel be set with one crane or more? If one, will mobile equipment be used, or will it be set with a tower crane? And what capacity is required to make all the lifts that will be encountered? How will block walls be constructed—with automatic mechanized scaffolding or with buck and board? The list is not complete until the supervisors have thought through construction of the entire project and listed, on paper, the equipment requirements for each operation.

When manpower levels are determined by craft, tool and supply needs can also be forecast. Obviously, everyone needs a hard hat, for example. With an allowance for loss and a projection of labor turnover during the course of the project, an accurate projection of the number of hard hats needed can be made. Knowing the forecast manpower level of pipe fitters and the duration of the mechanical phase of the job, an experienced piping superintendent can estimate the number of side grinders, end grinders, cut-off saws, and grinding discs needed, for example. Knowing the quantity of pipe and welding procedures called for in the specifications, the proper types and quantities of weld rod and wire needed can be projected.

Similar projections can be made for form lumber, plywood, nails, and form ties by carpenter supervisors. Forecast manpower levels used with the overall schedule are also used to determine the quantity and location of temporary facilities, such as portable toilets, craft sheds, gang boxes, and drinking water provisions.

PREVENTING SHORTAGES

The planning of a project in detail by the supervisors who will actually execute the construction will, in most cases, prevent equipment, tool, and supplies shortages from occurring during the execution stage. The most common reason for shortages, if the initial plan has been developed effectively, is the failure to assess the impact of changes in the plan on inventories. The second most common reason is underestimating or shaving estimated quantities in the mistaken belief that it will save money. Some managers believe that if they keep the quantity of a certain item in short supply, craftsmen will make more efficient use of the item. What usually happens instead is that the craftsmen hoard quantities of the item for use when the supply room runs out. Or, they resort to taking from each other what they need to get the job done.

On one reimbursable-cost project, the owner's representative arbitrarily decided not to approve the purchase of any more welding cable, notwithstanding the fact that late equipment deliveries had forced an increase in welders above the forecasted manpower level. He arrived at this position because he had, through personal checking, determined that the welding machines and their leads were only actually working—that is, depositing metal—about 45 percent of the time. His reasoning had caused him to conclude that there were twice as many welding machines on the job as were needed. He did not realize that because of the several other time-consuming tasks a welder performs—moving leads, moving into position, repositioning, replacing rod, and cleaning off slag—the 45 percent of the time that the machines were depositing metal was well above average, not below.

The resulting shortage of cable meant that when a welder had to make a weld at a point beyond the reach of his leads, he simply removed a section from a machine that was not making arc at the moment and connected it to his. A few moments later, when the welder whose leads had been taken got ready to continue his weld (which, in many instances, meant climbing up ladders or scaffolds, crawling along pipes or beams, and difficult positioning), he had no arc. The situation was chaotic until the project manager threatened to cut manpower and slow the job down unless the owner's representative approved additional purchases.

LABOR—THE REMAINING ELEMENT

The remaining element that must come to the point of the work at the right time and in the proper sequence is, of course, labor. But just placing labor at the work point, as one would a pile of bricks, and leaving it there for future use is not enough. The manager must recognize and understand the vast difference between inanimate elements, such as two-by-fours and bulldozers, and the craftsman who performs the labor. He cannot be treated like concrete block and steel—like a commodity. Concrete block will break if mishandled or overstressed; steel will bend; but the craftsman will react and spring back, most times in the wrong direction.

Installation, the act of putting everything together, is performed by a craftsman with thoughts and feelings about the situation in which he finds himself and reactions to the various outside influences to which he is exposed. He is therefore in a category by himself.

3

CRAFT LABOR

Chapter 2 discussed management's responsibility for bringing the plan, the tools and equipment, and the materials to the point of the work so the fourth vital quantity—labor—could do its job. But application of these elements must be in balance. The emphasis must be evenly applied.

Supervisors often tend to spend too much time optimizing equipment and tool applications and not enough time optimizing the application of labor; that is, making sure that the quality and quantity of labor used at the point of the work is correct or as close to correct as possible. They make sure that hoisting equipment is sized to lift the load and not oversized; they insist that the excavating machine for a small trench has a bucket width-squarely within an inch or so of the required trench width, and so on. But they do not spend enough time looking at the particular operation at hand and selecting the proper elements of labor to do the work. This is due, in part, to the technological advances the industry has made in the last several years. Today, the industry has computerized practically everything from timekeeping to maintenance programs. It has laser-beam surveying and computer-assisted design and drafting. The industry has optimized all of the other elements necessary to perform a particular field-construction operation, but it has just barely scratched the surface with labor.

PROBLEM-SOLVING TRAINING

Many analytical subjects, such as time-and-motion studies, are well-developed. This is because most of the managers in the construction industry have a technical background, and technical people like to deal with figures and make calculations. They have been taught the engineering problem-solving process of identifying the problem type, assembling the data, picking the correct formula, and solving the equation for the unknowns. They have also been taught that if they perform the above steps accurately, they will solve the problem. They have also been trained—their grades depended upon it—to believe that a problem has only one correct answer; the answer is either right, or it is wrong. They come out of school thoroughly proficient in this approach to probelm solving; many, in fact, believe that there is no other approach. This is the perfect attitude for anyone who does nothing but calculate values in mathematical equations, and it is a sought-after attitude by the manager who must have engineering solutions determined accurately.

SHADES OF GRAY

Understandably, these same engineers and technicians, when put into situations dealing with humans instead of steel, for example, tend to apply the same techniques and seek the same "right answer" each time. Unfortunately, solutions arrived at in this manner are rarely satisfactory. They see situations in monochromatic light: everything is black, or everything is white; there are no shades of gray. If they become managers without additional training, the results are usually less than adequate.

When dealing with labor, the engineer who has become a manager must recognize that the craftsman, unlike a piece of equipment, is a living, breathing creature, a reasoning entity. He must use more than the analytical techniques he was taught in engineering school; he must go a step further and look at the psychological impact on the craftsman of each decision he makes. He must assess this impact and predict the results. And the results are most likely going to be shades of gray, not the black or white he has been trained to see.

SELECTION OF LABOR

This chapter will look at many of the factors affecting the proper selection of labor and at the application of these factors at the point of the work. By looking at the makeup of labor itself, the manager can provide perhaps the single most important element of direct support to it. The manner in which a manager selects supervisors, the quantity and quality of skills required for a particular operation, and the crew size needed to execute that operation and then communicates his plan to everyone involved so that they understand the reasons and logic behind it will do as much to support the craftsman as will anything we have discussed so far.

Direct support of labor begins when the manager sets criteria for the project work force itself. He should define the needs he seeks to fill according to skill variety and experience levels, attitudes and behavioral patterns, timing and manpower levels. When he finishes, he should have a labor plan for the project that includes both quality and quantity.

STANDARDS OF ACCEPTANCE

Defining the labor quality sought for a given project begins when the manager, with an approximation of the craft requirements in mind, builds a profile of the craftsmen needed to execute the project.

To do this, he defines his or her company's basic standards of acceptance for the project. These standards should include details of the skill and experience required as well as any personal conduct standards and any special restrictions or conditions. Included also should be technical criteria developed from engineering specifications, such as tolerances on fit-up, welding procedures, the dimensional accuracies needed, and any other workmanship requirements.

Personal conduct policies should include behavioral and safety-related standards of acceptance. These should all be conveyed to the prospective employee when he is initially interviewed so that he will realize that they are conditions of his employment. He should be told of any special requirements, such as smoking restrictions, if any, and clothing standards. On many projects, especially those in existing facilities, plant rules already in effect are part of the contractor's contract, and therefore all personnel hired must adhere to them. Most of the time, this is for safety reasons, such as fire or explosion hazards.

It is much better, for example, that a craftsman who smokes finds out before he completes his application that he must go to a certain safe area if he wants to light up. This way, he can decide whether to accept the rule and hire on or reject it and go somewhere else to practice his skills. In some parts of the country, contractors require that personnel working outside wear long-sleeved shirts to prevent sunburn or hard-toed shoes to prevent foot injuries. If the craftsman is told about these rules beforehand and still goes to work, he accepts this condition of employment; if he finds out after he's already at work, he feels that somebody put something over on him. Consciously or unconsciously, he has negative feelings about that "somebody," and that "somebody"—his employer—is the one who should instead enjoy his highest respect and cooperation.

CRAFTSMAN-EMPLOYER COVENANT

The craftsman should be made aware of all the job rules relating to personal conduct—rules about absenteeism, on-the-job horseplay, drinking, and the use of drugs. He should also be made aware of the penalties involved in disregarding these rules. It should be made clear that the requirements so set out are standards of acceptance on the project; as such, they are conditions of employment. If the craftsman does not want to accept these rules as they are spelled out, he can choose not to work on that job. But if he accepts the job, he and his employer have entered into a covenant. The craftsman will apply his skills in accordance with the rules and any other conditions of employment; the employer will provide a safe and healthy workplace and pay the agreed-upon wages at the agreed-upon time.

Other standards of acceptance—those concerned with quality

and workmanship—relate directly to the craftsman's ego satisfaction and motivation after he is on the job. These also should be determined carefully and applied scrupulously.

CONSISTENT QUALITY OF LABOR

Just as a winning football team must have a good line and a good backfield to win football games, the disciplines and crafts on a construction project must be capable of achieving their goal: meeting the quality standards. The project must maintain consistent quality in its ironworkers, carpenters, and welders as well as in its electricians. It is the manager's job, working through his superintendents and foremen, to make sure that this aspect of direct support of the craftsman is thoroughly carried out. This is important because the ability of one craft to produce quality results depends, to a large extent, upon the results produced by the previous craft at that work point. Motivated craftsmen expect to produce quality results when they move in to a work area to start their work, and they have the right to expect that the crew before them did a good job in completing their part.

When a crew finds that the previous crew did not do their work properly, frustration and demotivation set in; they know that the defective work could have been prevented if management had been more diligent in assuring that its standards of acceptance were met. A high motivational climate will not exist if, for example, a crew of good ironworkers discovers, upon starting to set their steel, that the carpenters who set the footing anchor bolts missed their location by 2 or 3 inches. With today's fabrication and fit-up tolerances, if the anchor bolts are more than a quarter to a half inch off, remedial measures will probably have to be taken. If specification permits it, the column base plate may have to be slotted.Or the concrete around the bolts may have to be chipped out to a depth that will allow the bolts to be realigned correctly. If the error is too large, the footing will have to be removed altogether and replaced. In any case, the ironworkers cannot do what they do best—set steel—until the error is corrected.

In addition, since the production of a crew of workers depends on the skills of each of its individual members and their interaction with each other, it is vital that any craftsman hired, for example, as a journeyman indeed possess journeyman skills. Not only is this a simple production and safety consideration but it also is a critical factor in crew and individual motivation. Yet, in many instances, a craftsman is not properly evaluated before he is classified and put to work in a crew. The skill level he has mastered will become apparent to his fellow crewmen quickly, but by the time supervision finds out about the mis-classification and acts to correct it, days or weeks may have gone by with negative motivational results on the fully qualified craftsmen in the crew.

Consider the situation when this important requirement is ignored. Suppose two carpenters are hired as journeymen. They are paid the same rate, but one cannot build a concrete form that will safely withstand the pressures exerted by concrete placing operations, while the other constructs a form every time that contains the concrete true to line. Suppose, also, that the situation is tolerated by management and, further, that the craftsman who demonstrates the required skills and knowledge knows that management is aware of the other's shortcomings. One of two things will happen.

1. The proficient craftsman will conclude that management doesn't care about workmanship and will tolerate sloppy work, in which case he will relax his own personal standards.
2. He will not, in good conscience, be able to reduce his own standards of workmanship and will continue to build high-quality forms. The devastating feature of this alternative is that he will find some other project on which to work and build his forms. He will not stay.

The skilled craftsman does good work. But when he finds that his efforts and the demonstration of his skills are not needed—or at least not appreciated—he will lower his standards or he will go to a project on which the standards of acceptance are higher. Either way, the employer loses: if the craftsman stays, he will be demotivated; if he leaves, the employer will lose an asset—a highly skilled and motivated craftsman.

Besides general skills and knowledge, the manager should also seek, whenever possible, experience in the type of construction at hand.

SPECIFIC SKILLS

A preliminary determination of skills and experience is most easily made by reviewing the craftsman's job application. An alert personnel representative, by a few well-chosen questions in the interview, can assess an applicant's work experience and determine if the type of work he has been doing qualifies him for the job position that is being filled. The final determination, however, should be made by the craft superintendent or foreman, taking into account not only the craftsman's knowledge and craft skills but also the environment in which they are to be applied.

As an example, a supervisor requiring carpenters to build subsurface foundation concrete forms would not be well-advised to hire home-building carpenters who have made a career of finish-and-trim carpentry. Some of the basic tools look the same to the untrained eye, but even they differ from specialty to specialty. A claw hammer used to make forms, for example, has a different curvature and is 4 ounces heavier than one used to nail up house trim. Also tolerances, fit-up, size, weight, and actual material of installation are almost completely different. Both concrete form-making and house trim use wood as the primary material of construction, but that is as far as the similarity goes.

Moreover, trim carpenters get their job satisfaction from installing beautifully finished kiln-dried materials so that joints and fit-up are perfect. They know their work will be visible in someone's home for many years to come. In contrast, the form carpenter building a footing form uses heavy lumber and timber. He works to much lower tolerances—a quarter of an inch either way is usually acceptable. He knows that after the concrete is placed, he will be back to remove what he created only a day or so earlier. He derives his satisfaction from seeing that the form he constructed was strong enough to hold the wet concrete in place while it set up and strengthened; and that the process was carried out economically.

ENVIRONMENT OF THE WORKPLACE

Environment has an effect on the ability of a craftsman to perform his work efficiently. A pipe welder whose entire previous experience has been in a fabrication shop may not be the best selection for work 50 feet up in an open structure in cold, windy, noisy, and congested conditions. Pipe-shop welding is a manufacturing operation usually performed in a controlled environment with little outside distraction; field welding is an operation that requires intense concentration during the actual process. A welder with only shop experience may consciously or unconsciously be too affected by the noise, cold, or fearful height to maintain full concentration. For the same reason, most supervisors will not put a welder fresh out of training school in the same distracting situation. While the welder may pass the proficiency test easily, the test is conducted at ground level in an isolated booth. It takes usually a year or so for a recent graduate to become accustomed to all the situations encountered by a welder and to become fully qualified to weld safely on any job.

PROFICIENCY AND TASK LEVEL

It is equally important that the manager match the proficiency to the task. Some operations can be mastered completely in only a few repetitions; others take years to master. The manager should therefore look for craftsmen with not only the correct types of skills for a particular operation but with a proficiency level sufficient to perform the activity correctly the first time. At the same time, he should try to avoid overkill—hiring craftsmen with skills and experience far in excess of that required to do the assigned task.

When practical, he should assess the task complexity and take advantage of the opportunity to perform it with a crew chosen for its mix of skills and experience. More complex tasks should be performed by a more skilled crew and vice versa. As depicted in Figure 3-1, as task complexity increases from a_1 to a_2, crew

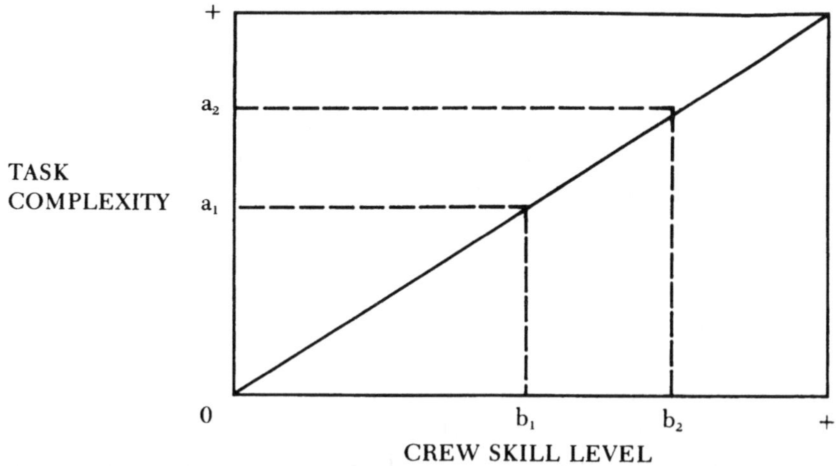

Figure 3-1 Crew Mix Skill Level vs. Task Complexity

skill level should increase from b_1 to b_2. Just as important is the converse. There are two reasons for this. One is cost. Less skilled or experienced craftsmen are paid at lower rates. There is no reason to use carpenters to remove forms if helpers and laborers can do it just as well. Second, a highly skilled and experienced craftsman is not challenged or motivated by work he finds subprofessional. In fact, he is demotivated by it. This doesn't mean that all skilled craftsmen are prima donnas. It simply means that the craftsman enjoys performing a task that is somewhat demanding of his skills. And when he enjoys what he's doing, he does it better and more efficiently.

For example, consider the activity of pulling heavy-duty, large-diameter electric cable from a substation through a utility bridge to a motor control center. Because most specifications require that the cable have no splices, it must be installed in one continuous piece. A long run can become quite heavy, requiring as many as ten to fifteen craftsmen to handle it. There is no reason for all of these craftsmen to be journeyman electricians. What is needed in this situation is more manpower but not necessarily higher skills. The only places in which journeyman skills are required are the beginning, the end, and the turn or bend points along the way, places where the cable could conceivably be damaged by sharp bending or chafing of the insulation if it is not carefully watched during the pulling process. The rest of the craftsmen, those doing the pulling, can be helpers or apprentices just learning to be electricians, those who need experience in performing such an operation and who can still find such an assignment challenging.

CREW MIX

In some types of work, the sequence of operation is repetitive and consistent; there is a mix of operations that requires high skills, medium skills, and low skills. For instance, in chemical and petrochemical plant work, a typical pipe crew consists of two welders, four pipe fitters, and two helpers. The pipe fitters bring the pipe together; line it up to acceptable tolerances; and orient the fittings, such as ells and tees, in the proper angles. All of this requires journeyman skills. Since most pipe is bulky and heavy, the helpers or apprentices assist the pipe fitters by supporting the pipe when needed and helping to turn and align it properly. When everything is lined up, the welders, whose craft requires a different skill, do the welding. Usually one welder can keep up with two fitters and one helper.

That's what crew mix is all about: selecting the proper mix of skills, experience levels, proficiencies, qualities, and quantities of the various types of labor available. Sometimes it is worthwhile to set up a special crew on an as-needed basis to perform a particular series of operations. At other times, the normal, or traditional, crew mix—a particular ratio of craftsmen to helpers that has evolved over the years in various types of work and classifications of projects—is sufficient. The manager's task is to determine which option to select in a given situation.

Crew sizes and mixes vary widely from job to job and from manager to manager. The physical layout of the work, the pace, the stage the job has reached, and the type of facility are all factors. Ratios that can be expected in the typical open-shop industrial plant construction are listed in Figure 3-2. In union construction, apprentice ratios are set by local or international agreements. Union-shop crews are about the same size as open shop crews for similar tasks, but usually there are fewer ap-

CRAFT	CREW SIZE	JOURNEYMEN	HELPERS
Carpenters	7	4	3
Millwrights	6	4	2
Ironworkers	7	6	1
Electricians	10	5	5
Cement Finishers	12	8	4
Pipe Fitters	8	6	2

Figure 3-2 Crew Size and Mix

prentices actually in the field, typically one or less per crew, averaged for all trades on the job.

When the mix of the crew is tailored to the operation at hand, everyone involved is challenged, doing what their level of training has prepared them for, and deriving ego satisfaction from the effort. This atmosphere encourages craftsmen to do more in order to enjoy more, and, in the process, they produce more units of work.

QUANTITY

Quantity of labor is as important as quality. As discussed in the previous chapter, the starting point for determining the best manpower levels is usually a careful review of the scope and estimate of the project to determine overall quantities of labor manhours required by basic activity. Developing this information further, and working with the schedule, the manager makes his own forecast of manpower levels in each craft for the entire project. (An ancillary advantage of doing it this way is that any divergences between the estimate and the manager's projections usually become apparent and can be resolved, or at least explained, immediately.) He follows this with detailed projections of the number of supervisors required to support the craft levels previously determined, and also the number of apprentices needed. He strives to design a smooth buildup of labor to a rounded peak and then an equally smooth reduction to the end of the project, as illustrated in Figure 3-3.

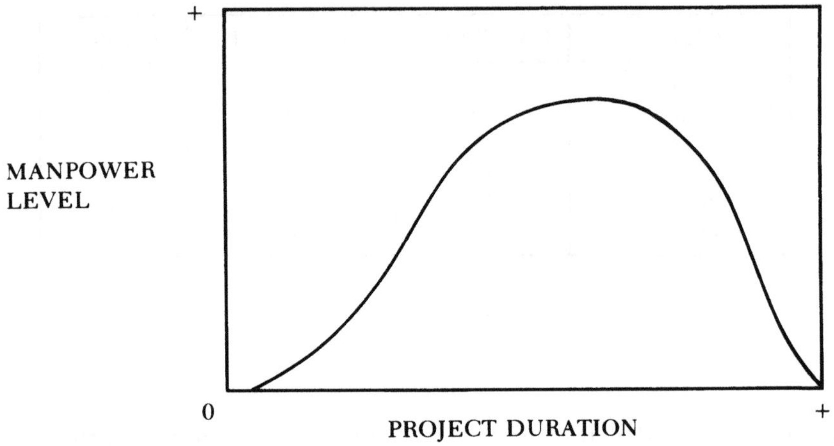

Figure 3-3 Typical Manpower Curve

ADDITIONAL INFLUENCING FACTORS

Many times when schedule considerations are the primary influences on a project, worried managers make panicked decisions that not only exacerbate the schedule problem but prove to be costly errors in judgment. One of these errors is increasing manpower density in a work area beyond that area's capacity to absorb the increase. This usually is done in an attempt to regain lost schedule time, but the result is just the opposite. Congestion (craftsmen physically getting in each other's way) actually delays completion of the work and causes productivity to fall, as illustrated in Figure 3-4.

Several years ago, a major U.S. corporation was building its largest manufacturing plant ever. On-time completion was vital to the company's financial plan because of year-end tax credits and, more important, the seasonal nature of the product to be manufactured. The entire first year's production had been presold. No slippage in schedule, therefore, could be tolerated.

Because of the size of the project, it commanded a very high profile in the company and attracted the continuing attention of top management. Unfortunately for the progress of the project, many of the members of top management had technical backgrounds and therefore a great deal of expertise in the process.

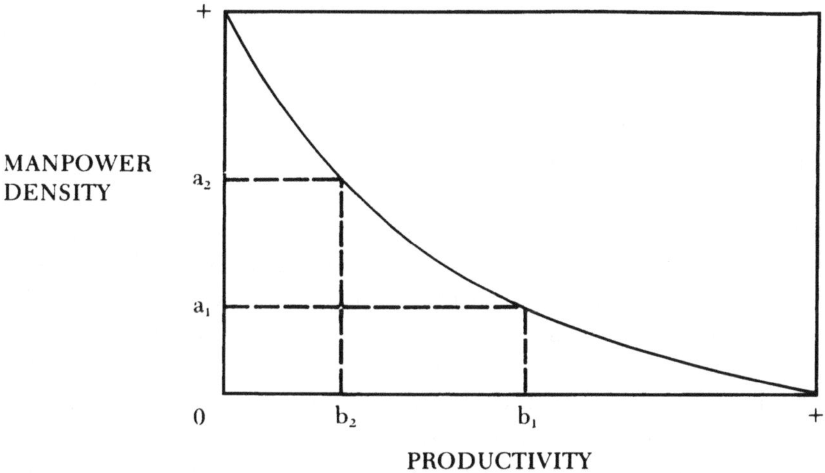

Figure 3-4 Manpower Density vs. Productivity

They could not refrain from diving into technical details and second-guessing the engineering team they had made responsible for the design. In their zeal to optimize the process and give everyone a chance to be heard, they caused decision making on several items of major equipment to be delayed past critical dates. The entire schedule, engineering and construction, suffered from the ripple effect of their indecision. Engineering immediately began slipping and consumed valuable slack in the schedule. Basic process design had been done, but *production engineering*—making the drawings and specifications that would be used in the field to construct the project—came to a halt. It depended on information from equipment manufacturers. Foundations could not be designed because no equipment weights and anchor-bolt layouts were available. Piping could not be sized because flow rates and flange locations and sizes were not available. Electrical engineering—design of the plant wiring and switch gear—was stymied because motor sizes had not been determined.

In turn, placement of other equipment, such as motor starters, transformers, and instrumentation was halted. This was particularly devastating because of the long manufacturing lead times traditionally required for electrical gear. Pipes, fittings, and valves could not be ordered, much less fabricated, for the same reasons.

Because of the late and, in some instances, revised decision making, forecast delivery of the critical equipment to the project site also slipped and consumed most of the remaining slack time in the schedule. The mandatory completion date set by the client's top management, however, remained unchanged.

When the necessary decisions were eventually made and firm dates for delivery of the essential equipment were obtained, a detailed construction-schedule revision was conducted by the contractor. This revision indicated that if there were no further delays and worker productivity remained at its then current level, the project would still be finished by the critical completion date. Also, the accompanying cost forecast indicated that the contractor's estimate would be met.

The owner's managers by this time realized what they had precipitated and became apprehensive that they would be criticized if the project didn't finish on schedule. After consultation among themselves, they directed the contractor to increase his pipe-fitter manpower by 50 percent on the remaining fieldwork.

The contractor protested this instruction and showed them his calculations that showed completion would be on time. He also pointed out the danger that costs would increase as a result of congestion in the work area and saturation of the contractor's field facilities. But finally, because of the terms of the contract, he had to comply with the instruction. The owner's construction representatives sincerely, but naively, believed that by increasing manpower by 50 percent, the remaining piping work would be finished in two-thirds of the time, an inverse proportion. They thought they could use what they calculated as the remaining third of the time as a sort of insurance policy.

The increase in manpower took place. As predicted, serious saturation and congestion took place. Where a manpower density of one man per 250 square feet would allow normal productivity in that type of operation, the added craftsmen caused the density to increase to approximately one man per 150 square feet. Productivity immediately dropped. The effect is illustrated in Figure 3-4. As manpower density increased from a_1 to a_2, productivity decreased from b_1 to b_2. Good craftsmen, frustrated by their inability to achieve their own personal daily goals, became discouraged and left the project. The quality of the workmanship decreased. *Rework*—corrections to improperly installed work—increased.

SATURATION AND CONGESTION

In the end, the increase in manpower yielded no improvement in the schedule at all. The extra third of the scheduled time the owner's representatives had counted on was eaten up by losses in productivity. Because of the congestion and saturation caused directly by the increase and the loss of motivation caused indirectly by it, *productivity*—lineal feet of pipe installed per man-hour—fell dramatically. Despite the heroics, the project actually finished in the same week that the contractor's projection without the increase in manpower had indicated it would.

The contractor's costs, however, were $1 million higher than they otherwise would have been. The loss of productivity due to the saturation and congestion was 50 percent. When everything was accounted for, the forced increase had caused the average unit installation cost of pipe to rise from one and one-half man-hours per lineal foot to over two and the cost of the last portion of the project to almost double.

UNDERSTAFFING VERSUS OVERSTAFFING

Every project has an optimum pace, one at which work areas are fully utilized, construction equipment is used efficiently, materials and supplies are readily available, and progress can be seen daily. With too little manpower, work suffers from a lack of momentum. The project's overhead expense, which is much more sensitive to time parameters than to manpower levels, increases simply because the project takes longer to complete. If, on the other hand, there is too much manpower, cost per unit of work produced increases because of *saturation and congestion*—too many people trying to occupy the same piece of ground or floor space at the same time and getting in each other's way—and because too much money must be spent renting or purchasing tools and equipment for a relatively short period of time. It is the manager's responsibility to set the job up so that an optimum pace is achieved and maintained.

On a jobwide basis, understaffing can slow the pace and drag the work out. Craft crews become lethargic. Remember that one of the reasons they entered the construction industry in the first place is the satisfaction they get from seeing some visible, physical thing take shape. They enjoy their craft best when they can see the results of their labor every day and every week. When they see many more opportunities for work than are being taken advantage of, they conclude that management lacks a sense of urgency in getting the job done. If management doesn't care about making progress, why should they? As a result, productivity suffers, motivation drops, and costs increase.

In addition, since a significant part of project overhead is time-related, overhead, which is an appreciable portion of total job cost, goes up. On a small job, project overhead may be as little as 40 percent of direct labor cost, while on a large stand-alone project, overhead may reach 100 percent of direct labor cost. *Direct labor cost* is the sum of wages paid to all craft labor performing installation of the permanent work; i.e., the total gross amount of their combined paychecks. Some of the more time-related indirect costs on a typical large project include management; salaried supervision; technical, administrative, and clerical staff; indirect craft labor, such as toolroom and warehouse attendants; construction equipment rental; office rental and services; and safety and security costs. It is not unusual for these costs to go up 50 percent or more on a job that is stretched out. This is in addition to the increase in direct labor cost due to the slower pace.

On the other hand, overstaffing can saturate project facilities and increase the hourly supervisory expense. In addition, congestion at work points can result in demotivation. Other indirect costs—including the costs of construction equipment, tools and supplies, craft sheds and gang boxes, parking, water distribution, and temporary toilets—can also go up.

The costs-versus-manpower relationship takes the form of a parabola. In Figure 3-5, the effect of too little manpower is illustrated by points a_1 and b_1; too much manpower by a_3 and b_3. At the optimum manpower level, cost, a_2, is at a minimum, b_2.

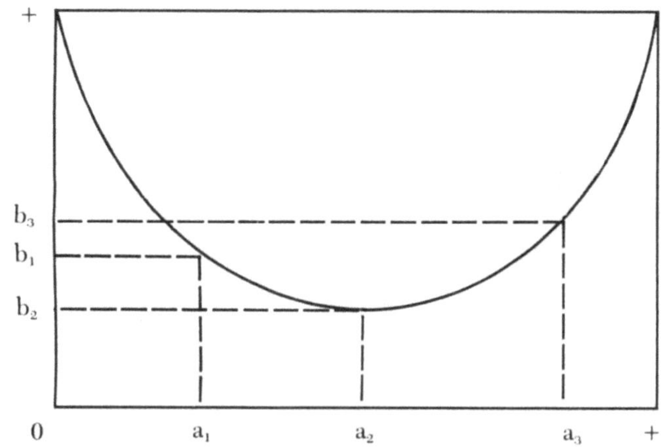

Figure 3-5 Cost vs. Manpower

MANPOWER LEVELING

Another way to minimize project cost and, at the same time, propagate a positive motivational climate is to avoid large peaks and valleys in the manpower curve as the project develops and is executed.

This is done by planning each operation carefully, making minor adjustments to the scheduled time of execution of each operation so that the peaks of the high spikes are shaved off and put into the adjacent valleys. Called *manpower leveling*, this is similar to the way a civil engineer designs the grade of a highway by balancing the earthwork cuts with the fills. Figure 3- 6 shows the difference between leveled and unleveled manpower curves. The optimum result is a gradual buildup in each craft to its maximum point and then a steady reduction in staff as the work ends. Desk-top computers and a plethora of sophisticated (by manual standards) scheduling and resource planning software now make it relatively easy for supervisors to make full use of each craft both while the work force is building up to its peak and while it is decreasing. The positive impact on craftsmen will be apparent throughout the project. They will not,

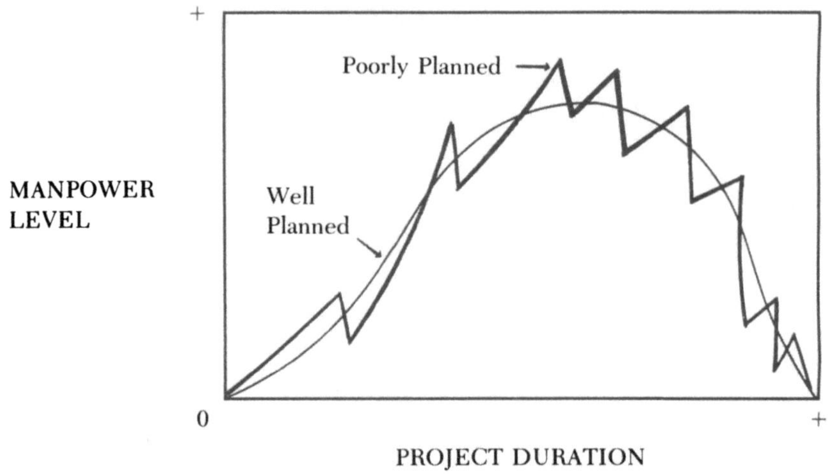

Figure 3-6 Manpower Leveling

consciously or unconsciously, slow down their effort as they perceive the work running out.

Experienced craftsmen know that one of the visible characteristics of a well-run job is a smooth buildup of manpower for each craft. They know it's possible because they've been on jobs where the staffing has been well-planned and coordinated with the other four vital elements described in the previous chapter.

They also know a poorly run job will show large peaks and valleys in manpower levels by craft. Typical of such a job is hiring, say, thirty carpenters one week, laying off fifteen the next, hiring twenty more the week after, and so on. This shortcoming in execution is usually due to faulty planning on the part of management, but sometimes an inexperienced manager will do it in an effort to nullify the impact of some delaying influence. It almost always costs more over the long run.

Actions of this sort have a tremendous demotivating effect on craftsmen, and the good ones will not tolerate it. They will leave as soon as they can and go to work on a project where the work is steadier. They know that better planning would have produced a smoother buildup and staff down of each craft. They'll go to a job where they can plan on being continuously employed until their part is finished. A stable work situation allows for more stability in their personal lives. It permits them, for example, to set up their car pools and get other logistical problems solved. This, then, minimizes the disruption in routine or rhythm mentioned by Maslow as part of the individual's safety needs.

A further effect of poorly planned labor staffing is that it requires much more hiring and laying off, bringing a continuous steam of new craftsmen to the project. This results in higher costs of hiring and lost time during the familiarization process.

Unlike some manufacturing situations, in which temporary layoffs are common and accepted by employees because they have a certain degree of job security (the auto industry is an example), in the construction industry, most craftsmen, once laid off from a project, do not return to it if they can avoid it. They reckon that the job is not well planned out and that all the other factors that cause frustration on the job will be present. To avoid this frustration, they go somewhere else to work, and when the first employer tries to hire them back, they are not available.

OVERTIME

Overtime is a two-edged sword: it can benefit a project if used judiciously and controlled closely; if not, it can kill it. Like many other management prerogatives, the negative motivational potential of overtime is greater than the positive motivational potential. Two categories of overtime exist: spot overtime and scheduled overtime. Of the two, spot overtime is somewhat less hazardous because it is limited to relatively predictable situations and does not invite abuse.

Spot Overtime

Spot overtime, as the name implies, is used in certain "spots" or situations to reduce labor cost. For instance, allowing a partial crew of cement finishers to work past their normal quitting time in order to trowel out a fresh concrete slab pour will allow concrete to be placed later in the day than would be possible were overtime not utilized. This means that more concrete placement can be scheduled for a given day. Bringing laborers in early to begin pumping out wet holes after an evening rainstorm may allow the main craft force to work a full day. Using a crew of electricians on a weekend to pull or install long lengths of cable may permit them to use otherwise busy roads or walkways to lay out and bundle the cable, thus expediting the operation. Application of spot overtime can be carefully planned and, on a well-managed project, will amount to less than 2 percent of the labor hours expended.

Its motivational value varies from neutral to mildly positive for those receiving it. Everyone is happy to get a little bit extra in his paycheck, especially if it doesn't happen so often that it's taken for granted.

Traps exist for the manager, however. One of these traps is impulsive or seat-of-the-pants consideration of overtime requests. To avoid this, the manager should develop a consistent set of cost-benefit criteria by which he analyzes all requests. Supervisors should be made aware of his criteria, and all subsequent requests should be compared against them. In instances in which the request does not meet the criteria, the requesting supervisor should be told why the request is being denied to prevent discouraging him from seeking cost-saving approvals in the future. In cases in which the request does meet the criteria, he should be praised.

Another trap is favoritism in making overtime assignments. Sometimes the same individuals or crews are given the available overtime assignments all or most of the time to the exclusion of others. Many times, this practice is unconscious on the supervisor's part, so care should be taken to distribute overtime assignments evenly among all employees who are qualified to perform the required tasks.

Scheduled Overtime

Scheduled overtime means working a portion or all of the work force extra hours for an extended period of time, usually a week or more and sometimes for an entire job. It is considerably more hazardous in terms of its potentially adverse effect on project-cost goals than is spot overtime. There are five reasons for scheduled overtime.

1. the need to clear bottlenecks in order to get ahead of schedule
2. plant shutdowns
3. catch-up plans to compensate for time lost due to inclement weather, late equipment deliveries, or late release of engineering
4. lack of labor resources
5. a deliberate slowdown by troublemakers

The first four reasons are legitimate reasons for overtime; the fifth is not. It is insidious and can spring from the other causes or be self-generated.

Clearing Bottlenecks

Scheduled overtime can help clear bottlenecks during construction. Used in the correct circumstances, it is a motivator because craftsmen see it as management's way of accomplishing some portion of the work that, until completed, will delay or prevent them from performing their part of the work. An example of this type would be doing the site grading on scheduled overtime so as to open up wider areas of the project for downstream operations, such as foundations. Another would be completing underground pipe and trench work to make parts of the project more accessible. A third might be putting roofing and siding operations on a fast-track basis to accomplish building dry-in before the cold or rainy season begins.

Plant Shutdowns

Managers in operating industrial plants put a price on lost production time that represents the price of the products that would have been sold had they been produced. These values can range from $10,000 to $50,000 per hour in a typical manufacturing plant. This means that when the plant must be taken out of service for repairs or maintenance or to make critical tie-ins of electrical or piping connections, it must be for an absolute minimum of time. Short projects might be scheduled at night or on a weekend. Longer ones, sometimes called *turnarounds*, are frequently scheduled around the clock.

These projects are positive motivators for craftsmen because they know that the duration of the scheduled overtime is strictly defined, the end is always in sight. And the extra pay they earn during the period makes the inconvenience of working long or odd hours well worth it. More important, many higher skilled and experienced craftsmen like shutdown work because of the challenge it presents: they must plan their work in great detail and with extremely close schedule accuracy, sometimes minute by minute. This leaves no latitude for planning error, and the craft work must be 100 percent correct the first time. Close teamwork is called for at every step along the way, and each craftsman depends on the previous craftsman to finish his work on time and true to the plans and specifications. Not every craftsman can function effectively under the pressure of this environment. But the sense of accomplishment the one who can gets by participating in a successful turnaround not only provides him with great ego satisfaction but also increases his self-confidence and self-esteem. If the achievement is in some way recognized by management—a simple "well done" works wonders—the motivational value of the experience is enhanced severalfold.

Catch-up Plans

Catch-up plans do not provide the same motivational boost as plant shutdowns. In fact, a long-drawn-out scheduled overtime plan—typically fifty-five hours or more a week—may have an inverse effect. The reason is that craftsmen know they have been let down by someone further back in the engineering-procurement-construction cycle. They know that some series of events has caused them to have to work extra hours. After the initial enjoyment of the extra pay wears off and they start to tire from the same routine day after day, they become resentful of those who caused the problem. Most of the time they do not know what the actual cause was, nor do they care: they blame it on management in general and the managers on their project in particular. About five or six weeks is all of this kind of overtime that can be imposed effectively; even then, productivity starts to drop after the second week or so. After that, craftsmen become so fatigued that they actually accomplish less in the extended week than they would have on a normal straight-time, forty-hour week. The work becomes less and less satisfying, and the end result is demotivation, which further reduces their productivity.

Lack of Labor Resources

When a project is to be built in a remote or sparsely populated area, skilled labor can be in short supply. It must be imported, and to attract qualified craftsmen, higher wages or longer hours must be offered. In union-shop work, travel allowances and paid travel time are the usual solution. In open-shop work, a standard workweek of up to fifty hours—five days at ten hours per day—is a common solution. In either case, the project is on scheduled overtime. Many contractors figure that productivity on a fifty-hour job is about 10 percent less than productivity on a normal forty-hour job, and they take this into account, along with premium pay—the extra "half" portion of the time-and-a-half wage mandated by federal law for any hours worked over forty in a week.

When it is explained at the beginning, the scheduled-overtime project attracts those who have already or feel they can work regular fifty-hour weeks. They accept it as a given when they hire on; as a result, there is no negative motivational impact. Most stay as long as they figure the extra pay they net after deduction of the expenses of gas, overnight lodging, and meals makes it worthwhile.

Then there are *camp jobs*, jobs in which the owner or the contractor sets up living facilities at or near the work site. Room, board, and recreational costs are paid in part by the employee and in part by the employer. As long as the workweek is held to fifty hours, "burnout" does not usually occur on these jobs. The camp itself can be a motivator or a demotivator, depending upon how it is designed and operated. Quality and quantity of food is the most important factor, with cleanliness and recreational facilities close seconds. One contractor, building a project on a remote site in the Rocky Mountains, built a camp with private rooms and bathrooms and a multipurpose recreational facility with game rooms, satellite television, and a beer lounge. He coupled this with first-class food—steak three times a week—tastefully prepared. Employees were charged about an hour's pay per day. Word of the living accommodations available on this project spread throughout the industry, and, as a result, the job, which peaked at over 3000 employees, was abundantly staffed with qualified craftsmen for its duration.

Another contractor approached a remote offshore project with a different concept. His idea was to work the employees so hard that they would be able to do nothing but eat, sleep, and work. He believed that the size of their paychecks was all that mattered to them and, accordingly, put the job on a seventy-hour schedule—seven days, ten hours per day—from the onset. His camp was poorly run, the food was mediocre, and "recreation" consisted of out-of-date magazines and one pool table for 700 residents.

The work schedule caused employee burnout, a high incidence of accidents, and heavy turnover—as much as 20 percent per month in some crafts. The demotivating effect of the poor camp conditions, coupled with the excessive workweek, destroyed any personal initiative in the craftsmen to do a good job. Craftsmen left as fast as their replacements arrived. They would come for a month or so—until they reached their earnings goal—and then leave. Schedule was met, but the loss in productivity caused job costs to be much higher than estimated.

Deliberate Slowdowns

The most hazardous condition a manager can face with regard to overtime is one in which a small group of troublemakers has talked the rest of the workers into slowing the project down by reducing their output, thus forcing a situation in which scheduled overtime becomes necessary. Any time a manager hears a worker ask, "When we goin' on 'time?" he needs to take action.

This question indicates two things: (1) there is a troublemaker in the work force and (2) this troublemaker believes that the scheduled completion date must be met at all costs, that the workers have management "over a barrel."

The condition is easy to spot by comparing actual unit labor costs with known standards or projections. A little investigative work will usually uncover the instigators. What happens next dictates whether the manager will lose control. If he takes swift and certain action with the nucleus and terminates the troublemakers from the project for some legitimate reason (these types usually break one or more job rules frequently), he can stop the movement from spreading. Other workers who might tend to try the same thing will perceive that they will be identified and similarly treated. If, on the other hand, he tries appeasement, ignoring the obvious ploy and exhorting the troublemakers to "try a little harder" or, worse, negotiating with them, he will immediately be recognized as weak, the criticalness of the schedule will be confirmed, and the situation will deteriorate. Either the completion date will be missed altogether, or it will be nominally met but at a greatly increased labor cost. The problem used to be common in union-shop work, but it also happens on some of the larger open-shop projects, especially in areas where the labor supply tends to be tight in certain crafts.

As with other instances of inappropriate conduct by a few, the good craftsmen—those who simply want to come to work each day and do their jobs—will be demotivated by a successful slowdown because they do not like to work at half speed. It destroys their rhythm. They also recognize that the slowdown indicates a weakness in management and thus expect other conditions affecting their freedom to use their skills without hindrance to follow. If the slowdown is dealt with firmly, on the other hand, they will recognize management's strength and rely on that strength to support them in doing their jobs.

As with other instances of inappropriate conduct by a few, the good craftsmen—those who simply want to put in a good [illegible] [illegible] [illegible] [illegible] [illegible] [illegible] [illegible] [illegible] [illegible] [illegible] [illegible] [illegible] [illegible] [illegible] [illegible] [illegible] [illegible] [illegible] getting them [illegible] [illegible] [illegible] [illegible] [illegible] [illegible] [illegible] [illegible] [illegible] [illegible]

4

Character of the Work Force

The previous chapter described many of the factors that affect the pace and flow of activity at the point of the work. When these factors are tended to properly by the support staff—supervisors, engineers, purchasing agents, warehousemen, planners, and others—progress is efficient because all the components are brought to the work point in a sequence and on a schedule that permit craftsmen to perform the labor of installation without delay. This direct support has a triple effect on craftsmen.

THE TRIPLE EFFECT

Greater Support Equals Greater Production

First, the pace of execution is steadier from one stage to the next. The better the preparation and coordination, the fewer the delays and the higher the productivity. Obviously, if craftsmen are not delayed, they can produce more work in a shorter period of time.

Management-Craftsman Covenant

The second, and not quite so apparent, effect this direct support has on craftsmen is a motivating one. When craftsmen see that management has done its part to get things ready for them, they look forward with enthusiasm to doing their part. The astute manager recognizes this second effect just as the good craftsman recognizes the first. Together, they form an unwritten covenant: when management supports craftsmen by getting the necessary elements to the work site, the craftsman reciprocates by performing the installation efficiently.

This is one of the most important covenants between labor and management. It is based on the concept that good craftsmen derive their greatest ego satisfaction from exercising their skills. For example, welders derive a great deal of enjoyment—ego satisfaction—from placing a good bead of weld metal. What they experience when their minds cooperate with their eyes and hands to accomplish a task is pleasurable to them. It is the essence of craftsmanship.

The craftsman's attitude can be summed up in the statement, "I get my satisfaction from being able to do my job without a lot of disruption and frustration. If management will get me the right plans, tools, and hardware, I'll do a good job and get the satisfaction I need." The manager's attitude can be summed up in the statement, "I've expended a lot of effort to prepare all the necessities at the point of the work for you, the craftsman, so you can perform the installation efficiently, without hindrance or frustration." When management does not perform its job—when the tools, machinery, equipment, materials, and planning don't reach the work point properly—the installation cannot proceed efficiently. Every time a craftsman has to stop work to go get a missing item, a crane or a box of nails, someone in management has not done the job properly. When this happens, craftsmen recognize it immediately; if it happens again and again, they adopt the attitude that "Management doesn't care enough about productivity to do the preparation necessary at the work point. Why should I care about producing efficiently for them?"

Projection

The third effect of direct support is another kind of motivational boost, one having to do with the intrinsic value that craftsmen assign to what they are doing and the resulting conclusion they reach regarding their own self-worth. Regardless of the size of a project, if it is perceived as successful or impressive by the public, or by others in the construction industry, craftsmen will derive a sense of satisfaction from participating in it. Psychologists call this *projection*; that is, by association or participation, one projects one's self into a much larger and more important accomplishment and feels entitled to a little of the credit for it. Just as the football team's guard or line backer is proud of having been a part of his team's victory in the Rose Bowl, craftsmen who bolted up the steel on an impressive new high-rise office building get a thrill every time they set eyes on the building.

They also derive satisfaction from knowing that there is a permanence about what they do. A testimony to their participation in a project will be there for them to remember, look back on, and perhaps even revisit. Craftsmen recognize that in many cases, a particular accomplishment or series of accomplishments will survive them and be there for their children and their children's children to see. This, in itself, gives them an enormous sense of satisfaction. The more noteworthy they perceive the project to be, the greater their pride in their association with it.

Craftsmen who worked on the restoration of the Statue of Liberty in 1985 are perhaps the best current example of this phenomenon. The utmost care was taken in every action involving the renovation of the 100-year-old "lady." Reports at the time indicated that there was an almost reverent attitude toward the work. The craftsmen were obviously proud to have been selected to work on the project; even more, they realized that their efforts would be there for all to see in another 100 years, something of which their descendants three or four generations hence could still be proud.

VISIBLE CONCERN

Still another benefit derived from direct support of the craftsman by management is the visible perception of concern. Simply by taking care of its end of the agreement regarding tools and machinery, management shows—makes visible for all to see—concern for the craftsmen and their psychological needs. The craftsmen realize that this concern by management makes it possible for them to do what they like best—exercise their skills.

As in many other aspects of the relationship between the

manager and the craftsman, the avoidance of the negative side is even more important. When management shows little evidence of concern for the craftsman, the craftsman interprets the signals quite accurately and concludes, "If they don't care about me, why should I care about them?" This leads to an indifferent attitude toward production, quality, and safety, with its attendant poor results.

It is important to note, also, that the concern must be genuine. It must be a constant, everyday action, a job-site habit by every supervisor and manager on the job if it is to work. The reason for this is that craftsmen will notice immediately any change in an otherwise poor management attitude. In such instances, craftsmen know that they only have to wait a short while to discover the reason for the change in attitude.

If, for example, a project site has been poorly maintained—there is trash all over, the field toilets are dirty, the equipment is full of mud, the work areas are dirty—and, suddenly, the manager takes a great interest in eliminating these unsightly and unhealthy conditions, the craftsmen will at once become suspicious. When someone from the home office or members of the local chamber of commerce come by for a job tour, craftsmen will know that the cleanup was for show and not out of concern for them or the conditions under which they have to work. Understandably enough, they will adopt a cynical attitude toward the entire process. When things return to their dismal normalcy the day after the grand tour, the craftsman's suspicions are justified and management's insincerity is confirmed, with predictable downstream negative results.

On one long-term maintenance contract, the craftsmen joked with each other that they could predict almost to the day when a VIP was scheduled to visit the site because all the normally dirty areas got a fresh coat of paint exactly a week in advance.

PROFILE OF THE CRAFTSMAN

Much of the material discussed thus far relates to the craftsman's ego, emotions, and reactions to the things he not only sees on a project but also feels or senses. To understand better just what motivates craftsmen and what demotivates them, it would be useful to take a look at the profile of today's craftsman. *Profile*, in this case, means a sketch of the typical craftsman that will be helpful in analyzing and predicting his behavior on the job.

The average craftsman is about 30 years old and has been in construction for about seven years. The average age of the construction worker has been decreasing over the past twenty years as those that received their construction training in the military during World War II and the Korean War reached retirement age.

The average craftsman has journeyman credentials. He has received formal training in his craft, in either a contractor-sponsored training program, a vocational training school, one of the military services, or a program conducted by one of the building trades unions.

He is proud of the skills he has acquired and enjoys applying them to new tasks. He likes to start new projects frequently to avoid the boredom sometimes associated with assembly-line or manufacturing jobs. Some of his work is art, he feels, because frequently a piece of work he produces instills in him a pride of accomplishment usually known only in the world of art.

One highly skilled welder, for example, carried with him at all times a wafer-sized piece of half-inch-diameter stainless steel pipe on which he had made a perfect butt weld as part of a test he had passed to become qualified to weld to nuclear procedure specifications. Another welder, who was certified in more than a half dozen welding procedures, enjoyed the creative aspects of his skill so much that he spent his evenings and weekends producing artistic weldments for his friends.

One master carpenter converted his garage at home into a fully outfitted woodworking shop where he made waterfoul decoys that were too perfect to be used by hunters. Instead, they were displayed as *objets d'art* in the homes of his friends—each one signed by the artist.

The average craftsman has completed twelve years of schooling, up from eleven years as recently as ten years ago. Almost all have graduated from high school and many have college training. A sampling on one project showed that 19 percent of the craft work force had some college level training and 3 percent had Associate or Bachelor degrees. In school, he was good at analytical subjects, such as mathematics and mechanical drawing.

He is a family man. He's married and has two children. He commutes from his home to his job daily, an average distance of 35 miles or one hour traveling time each way. About once a year, he takes an assignment that requires him to work farther from home; when he does, he's away from his family three or four nights a week.

He is buying a three-bedroom, bath-and-a-half home, drives a pickup truck, and owns a compact sedan for his wife to drive. He has the usual number of gadgets around the house, including a stereo, a VCR, a frost-free refrigerator and a programmable washing machine.

He works approximately 1850 hours a year, earning almost as much in wages as an engineering school graduate does in his first year out of school.

THE CRAFTSMAN AND THE ENGINEERING GRADUATE

Besides their annual earnings, there are other striking similarities between the craftsman and the recent engineering graduate. An analysis of some of these similarities may give the reader a better understanding of the personality of the craftsman and what he expects in his interpersonal relationships with his peers, supervisors, and managers. A comparison with the recent engineering graduate has been chosen as the vehicle with which to present these personality characteristics because many readers either have been at one time or will be in that category.

Having met the rigorous demands of an engineering degree program, the new graduate develops a deep sense of satisfaction, and rightly so. He is one of only approximately 5000 engineering students in the entire country graduating at that particular time. He has accomplished a significant goal in his life, one he may have set out for himself when he was half his current age. His sense of satisfaction comes from achieving the objective. He feels he has earned a reward and decides to buy himself some prizes as soon as he starts earning income. He may buy a new car, a new stereo, new clothes, or some other material possession that fills his need for gratification. He sets a pattern of challenge, achievement, and reward that he will follow throughout his entire professional career.

Sooner or later, the novelty of the rewards to which he treated himself wears off and he begins another iteration of the cycle. This time, the challenge is most likely on his job, and it entails putting the engineering skills he has acquired over a multiyear training program to work.

At this point, the quality of management he is exposed to determines whether or not he will be motivated to achieve. He no doubt recognizes the challenge if he's properly managed. He will be most motivated by tasks that cause him to stretch a little in a professional or intellectual sense, tasks that are not so simple he will become quickly bored but, at the same time, not so insurmountable that he will feel hopelessly inadequate. A good supervisor will calibrate the young engineer's basic ability—skill level—and rate of progress. He will give him increasingly more difficult challenges as his experience level increases. As he is allowed to pursue the objectives of the new challenge, he gains a sense of personal satisfaction out of each new accomplishment. Besides this, he hopes for or expects to be recognized also by three other groups for his achievements: his peers, his superiors, and his family.

He rates very high the probability of some kind of reward from his employer for successful goal accomplishment. While he would certainly like it, he realistically does not expect to receive a pay raise every time he accomplishes a short-term objective. However, he places a very high probability on receiving some other reward: praise, a pat on the back, or some other equally appropriate form of recognition. This is consistent with the Expectancy Theory.

The new graduate has gone to work for a company at a salary higher than he has been paid prior to becoming an engineer, probably more than he has ever earned before. His earnings are sufficient to provide the necessities and then some. He is not psychologically at either of the lower two levels of Maslow's hierarchy of needs. He is not concerned about food, clothing, or shelter; nor is he preoccupied with safety needs. He has the sense of security that comes from having a job and also from having the skills he has acquired in school. He feels they will be in demand in sufficient quantity for him to remain fully employed.

Instead, the engineer is at a level at which he feels the need to be needed. He needs to feel that he is an essential part of an organizational effort. He feels the need to be accepted by and to belong to that organization because of the feeling of relative self-worth it provides for him.

The craftsman has remarkably similar desires. He has been trained on the job, in a vocational-technical school, or both and has achieved his initial goal of learning the skills of his trade. The new challenges he sets for himself are directed toward achieving the standards of workmanship and knowledge that those in his craft are expected to possess. It is not surprising, then, to recognize that the craftsman has many of the same psychological needs as the new engineer.

He, too, is at a level at which he wants to feel needed by someone and needs to feel that he belongs to, and is an essential part of an organization or group engaged in achieving some worthy goal. He also is no longer at Maslow's survival level in the hierarchy of needs. In contrast to his predecessors twenty-five to fifty years ago, today's construction craftsman doesn't work simply to earn the money for clothing, food, and shelter. And more important from a psychological standpoint, he will not tolerate the abusive treatment that his father had to endure all through his working career. In comparison to those former times, there are plenty of jobs open to the good craftsman today.

EFFECT OF THE GREAT DEPRESSION

The great depression of the 1930s brought the country to a standstill and put millions of people out of work. There were hardly any jobs. People with a source of income felt fortunate to be able to do tasks far below their competence level. All walks of life were affected. Talented artists and sculptors, for example, were left without buyers for their art. They would do literally anything to earn four or five dollars a week. One ran a parking lot in a major city, earning twenty-five cents a day for each car parked.

Management practices suffered, and quality deteriorated. Untrained or uncaring supervisors in all industries heaped abuse on workers. They were unhappy times, and people took their anger and frustration out on those who could not retaliate.

The construction industry epitomized this national attitude. The craftsman of the period had no choice; if he wanted to feed his family, he had to put up with management under the "cussin'" approach described in Chapter 1. He had to take the abuse because he couldn't afford to get fired. He needed the job; he needed the money.

Today, any good craftsman can find work at a wage that will sustain him economically, one that will allow him to eat well and clothe his family. The craftsman's psychological needs, therefore, have escalated from the survival level to the belonging level, as have the young engineer's. They both want to belong and to be needed, to make a meaningful contribution.

The comparison between the craftsman and the new engineer can be taken further. They both want, and will work hard for, many of the same unnecessary material possessions mentioned in Chapter 1. This is due, in large part, to what they see on television.

TELEVISION—THE ELECTRONIC "WISH BOOK"

During the first half of this century, most Americans decided on their purchases of clothing, appliances, and similar goods after seeing them illustrated and advertised in printed media. Those in cities saw advertisements in the newspapers. Many people, however, lived away from metropolitan areas and got their ideas from publications of the great mail-order houses, such as Sears Roebuck and Montgomery Ward. Entire families would sit around the dining-room table in the evening, leafing for hours through the pages of such catalogs, dreaming about a new icebox, bicycle, or dress. "I wish I could have this" or "I wish I could have that" were often-heard phrases, hence the term, "wish book."

Today, although we can still thumb through mail-order catalogs, we have an even more convenient "wish book," an electronic one: television. It reaches virtually every household in every corner of the country. Families—rich and poor, young and old, large and small—are deluged with commercials that tempt them with ways to spend their money. Because of skillful advertising, those at every socioeconomic level are encouraged to want—and to believe they can possess—things that in earlier generations would have been obtainable only by those in higher income brackets.

The young engineering graduate sees a handsome, tanned actor playing a part in an advertisement for a seaside resort in Mexico or a ski resort in Colorado. He imagines himself in the scene. With the easy credit provided through one of his credit cards, he knows he can be there in twenty-four hours.

The craftsman sees the same advertisement and pictures himself on the beach or slopes, also. He, too, has credit cards and can put himself, literally, into that same scene if he wants to; and he does so with amazing regularity.

As a matter of fact, on a large refining project in western Colorado a few years ago, more craftsmen had ski racks on the roofs of their four-wheel-drive vehicles than had the more traditional gun racks in the rear windows of their pickup-trucks. A typical weekend at Aspen or Powderhorn found as many craftsmen as engineers and managers on the slopes.

The point is that media advertising appeals as much to the foreman, journeyman, and apprentice as it does to the manager, supervisor, and engineering graduate.

Television, the American electronic "wish book"; encourages everyone to want—and believe he or she can obtain—the same things. With such commonality of material wants and needs, it is not surprising that the craftsman and the manager also have similar psychological wants and needs.

EMPATHY BY THE SUPERVISOR

The real secret to a successful motivational program, and in turn to greater productivity, is the ability of the manager and supervisor to empathize with the craftsman.

The manager can visualize what will turn the craftsman on—and, more important, what will turn him off—simply by looking inwardly to himself. If the manager would be demotivated by a certain event, it is a good bet that the people working in the field as craftsmen would be, also.

Real estate experts say, only half-jokingly, that the three most important factors in locating a retail business are location, location, and location. In construction, the three most important factors in successful motivational management are empathy, empathy, and empathy; i.e., putting oneself in another's position and seeing things from the other's viewpoint. If the manager can empathize, he can set the motivational climate in which his employees will produce in abundance.

SATISFACTION EVERY DAY

One of the main reasons the craftsman has chosen to work in construction is because it enables him to see the results of his labor quickly and spectacularly. Each day, he can see what he has accomplished with his hands. And the daily cycle of exercising his skills and seeing its results is what he thrives on. He is happiest when he is productively exerting himself because he expects to see quickly the results of his effort and to derive his ego satisfaction from what he sees.

During the 1960s and 1970s, construction of nuclear power plants proliferated in the United States. Personnel managers found that the extremely long schedules—six to twelve years in many cases, as compared to the more normal six to eighteen months on other construction projects—made it necessary to rotate both managers and craftsmen after a period of time. The reason was the boredom and frustration that set in after a while. Physical progress was slow because of the myriad of safety and quality checks required, as well as the fundamental enormity of the task. Both managers and craftsmen lost interest in the project because they could not see the results of their labor as quickly as they were used to.

It's a basic fact, then, that the craftsman wants to perform his task so that he can obtain the satisfaction that having done a good job and made some visual progress brings. This is entirely consistent with McGregor's Theory Y. In sharp contrast to Theory X, the craftsman wants to work and will do so when properly supported by management and permitted to work in the right motivational climate.

CREATING THE CLIMATE—MANAGEMENT'S EIGHT BASIC RESPONSIBILITIES

Creating this climate is management's fundamental role. In addition to the elements of direct support described earlier, the proper motivational climate, when achieved, will be the result of a well-designed and skillfully executed comprehensive employee motivation program. Management has eight basic responsibilities in the program, and they will be discussed in the following chapters. They are.

1. to establish direct communication links
2. to provide consistent management
3. to remove sources of frustration
4. to fine tune the work force
5. to assure a safe working environment
6. to provide basic personal comforts
7. to provide training
8. to recognize achievement

5

Direct Communication

Communication can be thought of as the interactive transmission of signals between one person or group and another. *Signals* can be written or oral messages. They can also be understandings that are interpretations of actions—tonal inflections or physical projections, such as facial expressions. Many times, without adequate communication, one-way signals are perceived inaccurately and misunderstood by the receiver. In order to have efficient communication, one party must speak and one party must listen; they must then reverse roles. The more important of the two roles is, of course, listening.

LISTENERS ALWAYS GAIN

Truly attentive listeners always gain the most from a given communication episode because they know exactly what the other party has said and can make a rational decision regarding how to respond. In contrast, the poor listener, one who is composing his own next thought while the other party is talking, is at a distinct disadvantage because he usually does not know what his counterpart has said. In cases in which both parties are poor listeners, two separate and unrelated conversations often take place.

OPENING THE LINKS

On a construction project—where every activity involves people who need to know not only what the others are doing but also what they are thinking of doing—efficient, effective communication is vital. It is therefore important that the first positive steps in creating direct communication links between management and craftsmen start as early as possible. By starting the dialogue at the beginning of the project, when the overall motivational climate of the project is being created, a positive communication climate can be generated. In such an atmosphere, the manager can keep the day-to-day problems that inevitably come up on every project small—or even prevent them.

In opening the links early, the manager has an opportunity to prevent incorrect interpretation of signs and signals craftsmen may otherwise pick up. He can establish his credibility with the work force, which must be achieved before meaningful communication can begin. At the same time, he can assure the correct interpretation of his signals and reinforce his policies.

For many reasons, it is important for the manager to take the first step in opening a communication network with the work force. To begin with, the dialogue won't start at all unless the manager starts it. This is because the manager represents the source of craftsmen's paychecks, and, since there are almost as many managerial styles as there are managers, craftsmen are reluctant to make the first move for fear of inducing a negative reaction. Also, craftsmen will not, by nature, approach management just to talk. They are concerned that management may criticize them for spending too much time talking and not enough time working. They may feel inadequate when talking with someone at a high organizational level. Or maybe their own work ethic will not let them spend time in casual conversation while on the employer's time. They may approach the manager if a serious problem comes up, but not just to strike up an acquaintance.

CRAFTSMAN NETWORKS

More important, the manager must initiate the relationship so that word-of-mouth information about the project and its supervisors will be favorable and sufficient to attract good craftsmen. In an open-shop labor environment, most information about employment opportunities and job conditions is conveyed in this manner. Networking to exchange information is not something that has been passed down from the marketing and business arena to the work force. Craftsmen have been networking for fifty years or more. Many jobs are completely staffed in this manner. More than a few poorly managed projects are also stripped of their good craftsmen in the same manner.

On one very poorly managed project several years ago, the delivery of prefabricated pipe not only fell behind schedule because of inadequate expediting follow-up but also got out of phase with the required sequence of installation. As a result, the pipe fitters and welders, many of whom had been recruited just to do the installation, had very little to do for several weeks. Their supervisors, realizing that if they laid them off, they would almost certainly lose them permanently, simply staffed their crews with more craftsmen than required for the available work. In part, this was done also in the expectation that the pipe would arrive as each new week began. The result, however, was frustration among the craftsmen—the good craftsman wants to exercise his skill in order to achieve his daily measure of satisfaction—and many of them started communicating through their networks, primarily to find a project where things were better managed but also to tell their associates how bad things were on their present assignment.

Before long, word of a large new project buzzed through the job work force, and after the next payday, about two-thirds of the pipe fitters and welders quit and left for the new project. As a measure of how bad conditions were on the first project, the craftsmen left on the basis of rumor, not fact, that the new project would be hiring. Unfortunately, they traveled 800 miles to the new site only to find that, for several weeks, there would be jobs for no more than half of them. Rather than go back to the first job, however, they all remained in the area until the new project could absorb them. They did not know what conditions awaited them on the new project, but they knew those on the one they had just left. They reckoned that the new one could not be as bad as the old one.

BUILDING CREDIBILITY AND TRUST

The need to start communicating early cannot be overemphasized. It should be noted that at the beginning of a project, except for carry-over acquaintances from previous projects, the manager and the craftsmen are starting a new relationship. The manager wonders if the craftsmen are reliable and have a good work ethic, and the craftsmen wonder if the manager knows how to set up and operate a project efficiently. Unless the manager and at least a nucleus of the craftsmen have worked together successfully on a previous project, time must be dedicated to building credibility and mutual trust as the project begins.

At the beginning, the manager and the craftsmen will watch each other's actions closely. Each will draw conclusions from subtle signals that are given off. The objective of each participant is to try to determine if the other can help him achieve his goals. In reality, they are trying to find a way to help each other and, in turn, benefit themselves.

Not until both of them see the relationship as potentially wholesome will an effective communication link be established. When the manager is convinced that the craftsmen have the basic skills to accomplish the work, he will be ready to communicate. When the craftsmen reach the point at which they are willing to trust the manager, two-way communication can begin.

THE DANGER OF DELAYING

If a manager waits until a serious problem arises to begin a dialogue, he will have to solve a credibility problem with the work force in addition to the original problem. If he has, up until then, ignored the work force except to criticize it, if he has turned a deaf ear to its needs, if he has not taken the time to see things from its point of view, he will be received with skepticism when he tries to solve a problem that requires the close support and cooperation of the craftsmen.

In fact, he will have no credibility at all because the craftsmen will recognize at once what the situation really is and connect the manager's newfound interest in them to the problem that the manager is trying to solve. They will reject the manager's approach as openly self-serving. The relationship-building period must still be faced and successfully completed before the craftsmen will listen with belief to what the manager has to say. But the length of time it takes to get over the credibility hurdle is significantly longer and, in some cases, may be infinite. On a project that has gotten off on the wrong foot or has turned bad because of lack of communication or as a result of chronic and unresolved conflict between craftsmen and management, there may be no possibility of gaining the craftsmen's confidence at all—the manager has lost the game by default.

Consider the dilemma of the project manager who is suddenly faced with a union-representation election in one or more of the crafts on his project. How much acceptance can he expect if he has to approach a group of craftsmen for the first time, introduce himself, and present his reasons for voting against the union? He would have been far better off if he had taken the time to meet the craftsmen in the course of their day-to-day work on the project. He might be able to call some of them by their first names. If he had done this at the beginning, he might have been made aware of many, if not all, of the shortcomings or job problems that made the union appealing in the first place. He would have had an opportunity to eliminate these shortcomings as they were discovered, thus obviating the appeal of the union before it picked up momentum. The cost would have been far less, also. Repelling the thrust of a well-planned and well-financed union organizing campaign can be extremely expensive.

Opening a dialogue early on has other benefits for the manager. By taking the first step to get to know the craftsmen, the manager gains valuable insight into what signals the craftsmen are receiving and what they are thinking. In addition to learning about miscommunications, the manager will also learn firsthand, a lot about how the job itself is going from the craftsmen's point of view. This is a most valuable benefit because the information thus obtained is available to the manager through no other channels. By listening carefully to the craftsmen, he will learn what frustrations are really bothering them and may be able to mitigate these frustrations, as discussed in Chapter 7. In a more positive sense, the dialogue may help him identify areas of improvement, in work flow for example, that will improve productivity.

PROBLEM SOLVING AT THE CRAFTSMAN LEVEL

When the manager communicates successfully with the craftsmen, much of the frustration the craftsmen experience from seeing something wrong and feeling powerless to fix it is eliminated. This in turn improves productivity. Craftsmen want work to flow smoothly so that they can do their jobs effectively. They are also in a good position to spot problems and to devise solutions. They will offer their suggestions to the manager if the communication level is healthy.

Remember, no one is closer to the point of the work than the craftsman. No one can identify in detail the width and breadth of productivity-reducing problems as well as the craftsman can. There is no one better qualified than the craftsman to relate to management the presence of demotivators, such as poor work flow, and other sources of frustration.

OPEN-DOOR POLICY

A good communications program has several dimensions. Perhaps the most vital is a bona fide open-door policy. An open-door policy is a policy in which the manager's office door is easily accessible and kept open most or all of the time so that he can meet his employees and address their concerns. This is in contrast to a head-in-the-sand policy, characterized by the manager who keeps his door closed or blocks it with a maze of furniture and screens out potential conflict with a stone-faced secretary so he doesn't have to face these problems, at least not immediately.

An open-door policy means that if a craftsman has a gripe—real or simply perceived—and he's unable to get it resolved to his satisfaction with the foreman, he can talk with the general foreman and, in turn, the craft superintendent. And if he's unable to get it resolved with them, he can talk with the general superintendent or the project manager. The key is that he can keep going up through the organization until he feels that he has gotten a fair and impartial review.

In all instances, it is the craftsman's belief that he can voice his concerns without fear of retribution that makes the program a successful tool in the communication-improving effort. He must believe, by both the words management speaks and by the actions it has taken with previous cases, that he is safe and protected from reprisal during and after the process.

Many times, when face-to-face with the manager, lingering doubts will make the craftsman reluctant to talk about the real issues. He will talk around the problem and bring up secondary concerns, but he will have difficulty expressing what is really bothering him. In these cases, the manager must take the initiative in drawing out the real problem. He has to be able to detect when he has heard the fundamental problem and when he hasn't. It is a skill that the manager must develop early in his career if he is to be successful as a manager of people.

In a more positive sense, face-to-face meetings with the craftsman in a true open-door-policy atmosphere may yield some exceptionately good suggestions for improvements after the craftsman has gotten his gripes or whatever is bothering him off his chest. This is especially true if the craftsman's complaints have been heard out and the manager has committed himself to look into them. The manager may not accede to the request or promise absolutely to abate the gripe, but he should definitely agree to look objectively into the issue. Most times, this commitment is sufficient to satisfy the craftsman that his voice has been heard. Legitimate complaints, of course, must be acted upon if the policy is to be viewed by the craftsman as a real thing.

If nothing ever happens, no bad conditions are remedied, and no correctable shortcomings are acted upon, the manager will suffer a personal loss of respect and credibility and will be regarded as paying only lip service to the open-door concept. The project itself will be similarly regarded in negative terms as long as that particular manager remains in charge of it. Conversely, if the manager makes a genuine effort to back up his words with actions, many of the grievance-type problems that cause frustration among craftsmen can be prevented or eliminated. At the same time, frustration-induced problems, such as excessive absenteeism, high turnover, low productivity, or poor quality workmanship, may be eliminated.

UNWORKABLE PROBLEMS

Even if the manager is unable to solve the craftsman's problem or eliminate the source of his gripe, listening to the complaint, checking it out, and meeting with the craftsman again to relate what his investigation has uncovered will make the craftsman feel somewhat better simply because he has had an opportunity to talk directly to management. The craftsman can then decide whether to stay on the job and adapt to the problem or leave and find work on a project that does not present the same problem.

A remote project once had to operate an extensive housing program because of a lack of living space in the area. A single-status camp was built to house new employees initially. Staff housing administrators then canvassed the surrounding area in an effort to locate and lease family housing. These accommodations were assigned to employees on a first-come, first-serve basis as they became available.

Two craftsmen were told when they were recruited that family-status housing would be available for them and that the waiting period—the time they would have to live in the camp before their families could join them—would be about six weeks. Twelve weeks after starting work, they still had not been assigned family housing. With the project in a staff-up mode, the harried housing administrator, with more bodies than beds, simply was not able to fill the demand quickly enough. Housing was being leased, purchased, and even constructed as fast as possible, but still the waiting period stretched out—for everyone.

The two craftsmen, unable to get their needs satisfied at the housing administrator's level, next presented their problem to the project's administrative manager and, eventually, to the project manager. In each instance, their complaints were heard sympathetically (the project manager was in the same boat), and they were given consistent factual information. When they realized that management was trying to solve the problem and that they were being given the same treatment as everyone else, they accepted the situation. Largely influencing their acceptance was the fact that they had had what they felt were appropriate forums for their complaints. Management had taken the time to talk with them and explain the source of the difficulty—more employees needing family accommodations than had been anticipated. When they heard that the project manager had to wait even longer than they did, their dissatisfaction diminished.

Actually, a positive result grew out of the situation, even though the housing delay was not avoided. From the day they met, the two craftsmen and the project manager were on a first-name basis. The ice had been broken, the channels had been opened, and mutual trust and respect had been built. Much valuable information regarding other needs, suggestions for improvement, and messages the project manager wanted to get out to the craft work force informally were exchanged between them.

DIRECT CHANNELS

From this example, it should be clear that the manager should open the dialogue with the craftsmen directly and not through intermediaries. He has to do it himself and in person. He has to let the craftsmen see that he is a real person, not a fire-breathing monster, and that he is not going to intimidate them with his power as the boss. Admittedly, this puts a burden on the manager, but to be a successful leader, he must be willing to adjust his schedule so as to handle this vital need along with his regular work load.

The key is for him to get craftsmen feeling comfortable about seeing and talking freely with him about not only bad news but good news as well, all with the understanding that no reprisals will come to craftsmen who "say their piece" to the "big boss." The word gets around a job site very quickly if a craftsman is punished, directly or indirectly, for telling management what he thinks. The same job-site communications channels will also spread the good news if the opposite is true; that is, if the manager is listening to what craftsmen have to say.

It is virtually mandatory, therefore, that the manager reserve time each day to get out of his office and walk around the job, talking to the craftsmen he knows and meeting the new ones. By doing so, and just by showing his presence out where the work is going on, the manager sends a powerful signal to craftsmen that says "I'm interested in what you're doing. I'm interested in hearing about your problems so I can take action to solve them."

"WALKING THE JOB"

Townsend and Waterman, in "In Search of Excellence" (Harper & Row, 1982), called this concept "management by walking about." Many supervisors have managed this way for years without giving it a name. The key is that, instead of just observing the action taking place in the field, what is more important is for the manager to talk directly to the people that are making the action.

In addition to showing concern, the manager should point out that he's trying to create a win-win situation to benefit everyone involved in the project. When improvements are made and problems solved, the job runs smoothly. Under these conditions, the craftsman can exercise his skill, something he enjoys doing. He can thereby obtain the satisfaction and ego gratification that comes with having done a good job, whether that job be laying down a perfect bead or ringing a smooth trowel finish to a floor slab.

At the same time, the manager gets the project completed within the parameters of cost, safety, and quality that indicate to him and his superiors that he has done a good job of managing the project. Everybody wins, the optimal result.

There are several other benefits to be gained by the manager from "walking the job" and stopping to talk with craftsmen. The craftsman gets a boost to his ego as well as the increased esteem of his peers when the manager greets him by his first name and stops a minute just to say hello. Being greeted by the "big boss," who simply by his position in the project organization is in a position of respect, enhances the position of that particular craftsman in the eyes of his fellow workers. It also helps to create feelings of comfort and security on the home front when the craftsman can come home at night and tell his family that he spoke with the top man on the job that day. The family sees the boss as the provider of the craftsman's paycheck, and knowing that the craftsman is on a first-name basis with his boss adds to their sense of economic well-being.

GETTING STARTED

Building strong interpersonal relationships requires planning for most people. Close, direct communication links are often difficult for the manager to establish. To be successful, the manager may have to modify or add to his normal areas of interest for conversational purposes. For example, if he usually starts his morning by reading the business section of the newspaper, he might also spend a moment checking out the sports pages to see how the favorite local team is doing. If he is in an outdoor recreational area, perhaps he should acquaint himself with local hunting and fishing activity or the local stock-car races—whatever craftsmen on the project are likely to be interested in. If the manager can open the conversation with a topic of interest to the craftsman, he will have less difficulty in establishing meaningful communication in other areas.

In addition, the manager has to learn substantially more about the crafts with which he is not familiar. For instance, if his background has been primarily civil-structural, he should certainly learn the difference between TIG and MIG welding before talking to pipe fitters or welders. He may need to learn what a reverse thread right coupling is before talking with plumbers. He needs to know something about the language of the craft. This preparation by the manager will enable him to break the ice and at least start a dialogue. When he begins a meeting, he does not know exactly what subject—beyond the weather, perhaps—will come up, but one probable topic could be the craftsman's craft. If the craftsman is good at it, he will enjoy talking about it; and so this is the perfect spot for a manager to begin a conversation.

Once a somewhat relaxed dialogue has begun, the craftsman will feel freer to make suggestions. The manager will find he is able, in many cases, to make direct improvements in productivity by making changes suggested by craftsmen. They will make these suggestions if they feel management is interested enough to listen and to take the suggestions seriously. If management takes action on the suggestions that have merit, the effect on the work force will be dramatic. The craftsman making the suggestion will think, "I did something with my head as well as my hands; they listened to what I had to say and agreed with me. Hopefully, they will see me as a valuable employee and recognize my contribution. In any case, I have the self-satisfaction that comes from having an idea of mine accepted and used by management."

UNCOVERING DISCONTENT

In addition, when communications links are opened early, the manager will find that craftsmen will be more open with him later when it may be necessary to discuss important things, such as wages, benefits, and working conditions. He will also be able to uncover sources of discontent through everyday, casual conversation and to take action to eliminate those sources before the discontent becomes widespread or severe.

A contractor once opened up a pipe-fabrication shop and staffed it with about fifty craftsmen. The shop was a rather uncomplicated operation, the contractor thought, and so, making what turned out to be a poor decision, he let an out-of-work tennis partner manage it instead of hiring someone with strong management skills or, at least, experience in pipe fabrication. The incident that followed was therefore caused by the company owner himself, i.e., by top management. This is important to understand because it illustrates how each level of management—even the top level—must be watchful for adverse consequences in the decisions they make, especially when those decisions relate directly to people.

Because of the new shop manager's lack of skills, the operation, although it had the finest equipment available and was staffed with top-notch craftsmen, did not turn a profit. In an attempt to improve the situation, and without doing much in the way of problem analysis, the manager assumed that the cause of the poor performance was the craftsmen. He decided to rout out the "dead wood" in the shop. To determine exactly who was to be fired, he installed a see-through mirror in the wall between his office and the shop so he could view the entire floor without being seen himself.

The first victim was a young craftsman whom the manager observed one day making what he felt were too many calls from a pay telephone on one wall of the shop. Without inquiring further, he went out on the floor and, in full view of everyone, fired the offending craftsman.

Rather than being the end of the manager's problems, it was the beginning. If he had checked, he would have found that the craftsman, who was new to the area, was calling his pregnant wife, who was due to give birth to their first child any moment. He simply wanted to make sure she was alright. The couple had talked it over and agreed that since the shop was only a short distance from where they lived, he would go in to work, and check with her a couple of times during the day. If she was having any difficulties, he would be able to get home quickly. The irony of the situation was that the craftsman was so dedicated to his job that he was making the calls while the others were taking breaks so that he wouldn't hold his fellow workers up.

The shop manager, even when he learned what the situation was, refused to back down and reinstate the craftsman. Obviously, his own ego got in the way, and he couldn't bear to admit he had been wrong.

The story quickly spread to other craftsmen in the area, and it wasn't long before an organizer from one of the national building trades unions was sent to town to evaluate the opportunity presented by the occurrence. After talking quietly and away from the shop with many of the craftsmen, the union organizer realized that the incident was only the tip of an iceberg and that a high level of worker discontent existed at the shop for many reasons. Seizing on the opportunity presented by the widespread dissatisfaction, the union dedicated funds from its war chest and began a full-blown organizing campaign. The aim of the union was to recruit not only the fifty craftsmen working in this particular shop but, eventually, all the contractor's other workers, who numbered in the thousands. The union figured that getting a toehold in the shop would enhance their chances of organizing the rest of the company. The requisite number of signatures was obtained, and an election was ordered by the National Labor Relations Board.

The manager, now fearful for his own job, awkwardly swallowed a bit of his ego and tried to open a dialogue with craftsmen in the shop. One can imagine what the response was. He tried to mount a campaign to win the election. He tried to get the craftsmen to listen to his thoughts regarding the disadvantages of unions and the benefits of remaining an open shop. But the fact that he had never made an attempt to communicate with the craftsmen before put him in the position of talking to strangers. He had never even spoken to most of the craftsmen. They knew him only by the signs and signals he had put out, the most damaging of which was the see-through mirror. The craftsmen correctly concluded that the manager did not trust them to work productively and, in return, they did not put any faith or trust in him.

It was not until the company owner found out about the debacle, came to the craftsmen (many of whom he knew by name from the days when both he and the company were just starting out), and got a firsthand grasp of the situation that the tide turned. Because of the communications links and mutual trust that had been previously built, the craftsmen believed the owner when he talked to them.

In the election that followed the employees decided not to have the union represent them. The manager who had caused the problem was replaced by the production manager, himself a craftsman who had worked his way up through the ranks. The new manager knew the value of communicating with the work force because he had been part of it for years. He had no trouble maintaining the lines of communication he had opened previously and gaining the trust of the workers. The atmosphere improved in the shop, and productivity increased. The see-through mirror was removed, and the fired craftsman was reinstated.

The value of getting communication lines open early and effectively cannot be overemphasized. Without them, the manager is really in the dark. He has nothing but reports and printouts with which to monitor the project, and the information contained in this material is frequently shaded and filtered. The manager who relies on it alone is effectively insulated. His situation is similar to that of a doctor trying to listen to a patient's heart with a radio blasting nearby—there are too many extraneous inputs jamming his senses for him to reach a correct conclusion. He can try textbook approaches to solving the problems of missed schedules, overrun budgets, and poor quality, but until he gets to a level of dialogue sufficiently open and free-flowing in both directions, he is effectively prevented from getting to the root causes when things do not go right. He is virtually powerless to implement corrective action because he does not have the advantage of knowing what the craftsmen think or how they might be brought into closer coordination with each other and management.

The manager must, then, after opening the dialogue, be on the lookout for ways to broaden the communications loop. The open-door policy and face-to-face conversation techniques form the foundation for several other methods of expanding the channels of communication between management and the labor force. They fall into two general categories: one-way communication and two-way communication.

One-Way Communication

One-way management communication devices include bulletin boards, paycheck inserts, newsletters, and, on larger jobs, project periodical publications. All of these devices allow management to say something to the work force. They are good for passing on information for which a response is not really required. Information transmitted through these devices include safety awareness tips, work rules, working hours, holidays, important phone numbers, and so on. They may also include recognition items, such as which crew had the best attendance record, highest increase in productivity, or greatest number of accident-free hours.

One-way craftsman communication devices include suggestion boxes and telephone hot lines. These devices allow the craftsman to say something to management, anonymously, if he wishes. He can jot it down on a piece of paper and drop it in a suggestion box, or he can call a specially designated phone number and leave his message on an answering machine. Either way, he has an opportunity to say something to management and, if it is a gripe, get it off his chest.

His opinion of management will be either improved or diminished depending on the action it implements. The Expectancy Theory explains some of the reactions. The craftsman may complain about wages if he feels one particular craft is getting too much as compared to another or if he feels he, himself, should be paid more. But if, in fact, he is getting as much as others with the same level of skills, then, although he may gripe about it, he really expects to be paid about what he is getting. Having a means to convey what is on his mind directly to management is, in itself, a motivator because he knows that management has at least heard him.

All one-way communications from the craft work force should be acknowledged by management. Usually, the simplest way to do this is to quote a few key words or the central theme of the message in a short note on the bulletin board or in the newsletter, along with a comment that the suggestion has been received and will be considered or acted upon. Updates on any action in progress are important just to let the craftsman know that his idea has not been forgotten or lost in the red tape that pervades many organizations.

It is frequently advisable to acknowledge even attempts at humor. The key is for management to let craftsmen know, somehow, that their voices have been heard. Some exceptions apply, however. There is always the possibility of abuse, especially with answering machines, and obscene phone calls should not be acknowledged in any way whatsoever. It is rarely possible to punish those responsible for these phone calls, so the remaining option is to recognize, through praise and positive reinforcement, those who make good or meaningful suggestions.

One project manager, anticipating the potential for abuse, decided to head off any nasty phone calls by putting one of the young office secretaries in charge of the program. She was charged with the task of composing and recording a pleasant outgoing message and transcribing any incoming messages on the tape each morning. When he announced the implementation of the hot line, the manager made sure everyone knew that this particular secretary would be the one listening to any messages and requested that users take her youth into account in both the language and content of their calls.

He also revealed that the young lady was the daughter of one of the largest ironworker foremen on the project. Needless to say, the hot line was an immense success; most callers with good suggestions left their names—young bachelors, perhaps, more than the rest—and even the gripes and attempts at humor were in good taste.

Two-Way Communication

Besides "walking the job" and talking to various craftsmen, there are other ways for the alert manager to stimulate a dialogue. Opportunities need not be limited to job-site situations. In fact, some activities, such as team sports and similar recreational activities, provide an excellent arena because craftsmen recognize that these activities take place on "neutral ground," places where a person's position on the job site is disregarded in favor of his effort and ability in the particular sports activity. Company sponsorship of a team in a city softball or bowling league is inexpensive compared to the opportunity it provides for communication in a relatively relaxed atmosphere. On a large project, interest may be strong enough for the company to sponsor an entire league.

On one project, a craft superintendent volunteered to organize and manage a softball league in which the teams would be organized along craft lines. The piping department team, for example, would be open to anyone involved, in some way, with the pipe work on the project. In addition to the obvious job categories, such as fitters and welders, there were the engineers and draftsmen who detailed pipe layouts, the purchasing agents and buyers who procured it, the warehousemen who stored it, and the crane operators who hoisted it. Managers and supervisors would be invited to join any team that fell within their scope of responsibility.

In some cases, the rules regarding craft association were stretched a bit to allow recruiting otherwise ineligible players to join a certain team. In one memorable instance, a security guard was permitted to join the millwright department team because he had issued the gate pass for a truckload of heavy machinery when it arrived on site. (It was later learned that the team manager's interest in him had intensified when he learned that the guard had once been a professional baseball player.)

The league created an environment in which the various disciplines and levels of specialty could get to know each other. Back on the job site, this acquaintance enabled the workers to work with each other more efficiently. Whereas once a foreman might not have been too concerned about expediting the release of a crane, for example, the simple fact that he knew the supervisor of the workers needing the machine next induced him to finish a little earlier so as not to delay them.

The managers who played enjoyed even greater benefits. Not only did they get on a first-name basis with many of the craftsmen more quickly, but they had something to talk to them about back on the job. Identifying discontent and problems as well as soliciting suggestions for improvement became much easier because of the relationships that had been initiated on the ball field.

BEST-CREW AWARDS

Other forums can be created and sponsored by the company on the job site itself. If structured and used properly, they are to everyone's benefit. Sincerity and fairness are the key. One of these forums—especially effective because it both creates a healthy climate for dialogue and also recognizes achievement—is the best-crew luncheon.

There are several categories of recognition for crew performance. They include "best improvement in performance," in which the crew that performs best against the labor control estimate is recognized. Another is the "crew with the best improvement in safety record," based on a combination of parameters, usually including least number of observed unsafe acts, fewest observed unsafe conditions, least number of first-aid and doctor cases, and so on. Still a third best-crew award is recognition for "best work attendance," as measured by lowest absenteeism.

The program is inexpensive to create and operate and, when executed properly, provides a wealth of useful, often vital information on which management can act to improve the health of the project. Craftsmen enjoy the recognition and the opportunity to talk to management. Operationally, it is simple to implement.

By a process acknowledged to be impartial by everyone involved, the best crew in each category is selected, usually on a monthly basis. In addition to individual certificates and visual recognition symbols, such as decals or medallions for their hard hats, the best crew is invited to a catered lunch with top management, usually including the general superintendent and the project manager. During the lunch, the project manager demonstrates his interest in the craftsmen and congratulates them on their achievement. Conversation at the lunch may include topics such as the particular plans or actions the crew used to win the top slot in the competition.

The manager should show a personal interest in the crew. He might inquire about their families, their children, the kind of sports they follow, and their interest in fishing or hunting.

Once the crew is comfortable (remember, while they may be comfortable chatting with the boss as he walks past them out in the field, they are not used to sitting down to lunch with him, on his "turf," and it takes a few minutes for them to relax), the manager can inquire about conditions on the job from their point of view. Are the field toilets cleaned out often enough? Is the drinking water fresh and cool? Is it available at all work locations throughout the day? Is there a problem with dust in the parking lot? Is the procedure for drawing out tools too cumbersome? Does it take too long to obtain supplies from the supply room? Is the flow of materials efficient? What can be done to make their jobs run more smoothly?

The types of information to be derived in this type of meeting will usually not deal with the shortcomings of people. The craftsman cannot be expected to comment on the quality of his supervision or on other craftsmen in an open group session. However he may be quite open, after he is made to feel comfortable, in pointing out and talking about job conditions or practices that are frustrating or that need to be improved.

For this type of dialogue to be successful, the manager must do most of the work. He must learn the craftsmen's names if he has not previously done so. He must pronounce them correctly and without hesitation. Little things mean a lot in this situation. He must have a working familiarity with their assignments and the part of the job they are working on. He must know of any special conditions the crew is enduring, such as working in wet holes or high above the ground. In other words, he must do his homework before he walks into the meeting. Detailed knowledge and genuine concern will surprise most craftsmen, who feel that management doesn't even know they exist, much less what they are doing.

TAKING ACTION

If the channels opened up by best-crew awards and similar activities are to continue to be a source of information for the manager, he must be willing and able to take action on the problems that can be corrected. In other words, he must deliver. Craftsmen must get feedback on their comments and suggestions or they will stop offering them. Moreover, when management takes action, productivity improves, not only because the project manager has removed the cause of the delay in the toolroom by assigning more attendants or moved the toolroom closer to the work area to reduce travel time, but also because craftsmen, seeing the manager's actions, know that management has listened and heeded their suggestions. This, in turn, gives them a feeling of self-worth and self-esteem and a perception of increased esteem from their peers. They understand that management has recognized them for both their achievement and their suggestions, and they will strive to gain more of this recognition, because they want the ego satisfaction that comes with it.

NEWSLETTERS

Another dual-purpose device is the periodic newsletter sent to the craftsman's home. This is a good communication device not only because it recognizes the craftsman's achievements in the presence of those whose esteem he values the most—his family—but also because it allows management to convey certain thoughts to the craftsman in an environment where they may receive a more relaxed and considered acceptance.

Many managers try to forget about the project when they are at home, but the craftsman who is enjoying his job on a particular project is proud of his accomplishments and pleased to receive a communication that describes some feature of his project and his part in it. He gets a great deal of satisfaction out of showing his family the pictures and articles in each issue, and if his name is mentioned, he gets an extra boost.

In summary, direct communication is vital to a successful relationship among the people working on a project. Free communication up and down the organization as well as across it from department to department will encourage better understanding and cause individuals to cooperate better. Direct communication between the manager and the craftsman will allow blockages to productivity to be removed and will generate a more wholesome atmosphere in which craftsmen can not only perform their craft but enjoy doing it. This, in turn, creates the necessary healthy climate for craftsmen to motivate themselves to do an even better job.

6

Consistent Management

The fact that craftsmen are at or aspire to Maslow's belongingness and love needs level (Chapter 1) does not mean that external influences, such as mismanagement by supervisors, can't drive them down to the next lower level or below, with a proportional loss of motivation.

EFFECTS OF POOR MANAGEMENT

Poor management can cause craftsmen to lose their sense of security, the presence of which permitted them to reach the higher level in the first place. Similarly, unexpected changes and the loss of things the craftsman has come to expect each day can cause him to lose his motivation. What he looks for in his working environment is *consistent management,* the establishment and even application of policies, procedures, and standards of acceptance by supervisors and managers.

Management thus has an obligation to develop or formulate rules to govern the actions and conduct of all personnel on the job. Management has the obligation also of communicating these rules clearly and effectively to the work force, a responsibility that was amplified in the previous chapter.

In carrying out this responsibility, the most important points to remember are (1) that rules so developed should be realistic and well thought out and (2) that they should be applied evenly to everyone on the project.

Realistic Rules

The reason for the first point should be obvious. As part of the hiring-in covenant, the craftsman agreed to comply with the rules. In general, he will go along with any realistic extrapolation of them as long as it is consistent with what was in effect when he was hired. But he will not tolerate (for any longer than it takes to find employment elsewhere) rules that are either legally or practically unenforceable; rules that because of their inherent weakness or lack of wisdom, must be continually changed; or rules that are whimsical products of egotistical managers.

As often happens in a family business, a father once promoted his son much too quickly for the son's own good. He made him administrative vice president after less than a year in the business. Not only was the son in a situation far beyond his competence level, but, even worse, he deluded himself into believing he deserved the promotion because of his past performance, which, in reality, had been mediocre at best. He decided to try out his newly bestowed power by taking administrative control and whipping the company into shape.

Wanting to test his importance and consolidate his position but reluctant to tackle anything that carried with it the risks of failure or conflict for which he was not prepared, he attempted to devise an initial action that would make it easy for him to assert his newly acquired power. He decided to expedite the flow of paperwork by posting a notice that each employee was to complete his time card and get it in by 10 A.M. of the last day of the week, not in itself an unrealistic request. However, he followed this up by stating that any employee failing to comply would be penalized by having his pay withheld until the next pay period. The action, of course, backfired. The notice quickly became a laughing matter because most of the employees knew that it was contrary to the labor code in that particular state for an employer to try to coerce employees by threatening to withhold their pay.

Unfortunately, the matter was also a tragedy because it demotivated the entire employee group. People asked themselves and each other, "If Junior can't keep something as simple as salary administration rules and regulations straight, what will happen when projects go into high gear and people have to depend on each other to know their jobs and perform them correctly the first time?" Not only did the son not gain the immediate respect of people working in the company, but he actually killed any chance of ever gaining it.

If an employee is chronically tardy in turning in a time sheet or any other information on which others depend and there is no plausible reason for it, he should be counseled or reprimanded. In many cases, such an employee will be found to have other shortcomings and should be carefully evaluated regarding his continued employment.

Even Application

Craftsmen recognize the need for and, in fact, welcome job rules and job discipline. They make the explicit decision to accept the rules when they agree to go to work. At that time, if they do not want to wear safety glasses or hard-toed shoes, if they prefer not to open their lunch boxes for inspection when exiting the project, or if they want to be able to smoke whenever or wherever they wish, then they can choose not to accept employment. If they want to comply, they sign on; if they do not, they go elsewhere to find work.

If they do sign on, then they have good reason to expect that the rules and standards that have been outlined to them will be applied evenly to everyone on the project, up and down the organization. There may be aspects of the rules that they don't like, but if everyone else has to conform to them, they feel, "I don't like it, but if everybody else has to do the same, I guess I can, too." They will comply until they become aware that someone else is not doing the same. Then they feel exploited, taken advantage of, and discriminated against. They also feel that either management changed the rules after the deal was made, doesn't have the fortitude to enforce the rules, or has deceived them. In any case, management loses its credibility. The result of all this, of course, is to decrease the motivational level.

RANK AND PRIVILEGES

The pre-World War II attitude in the construction industry that "rank has its privileges" is retiring with the veterans that propagated it. The manager who expects to create and maintain a positive motivational climate must look carefully for remaining vestiges of differential treatment of employees. Beyond the perquisites awarded on the basis of individual achievement, as described in Chapter 12, this attitude has little value in the workplace today.

It is vital to the motivational health of the project organization that the "privileges-of-rank" syndrome be avoided for the obvious reason that the "haves" feel and act unjustifiably superior, and the "have-nots" feel, just as unjustifiably, inferior. It conjures up in craftsmen perceptions of class discrimination, may promote aggressive reactive behavior and negative feelings, and always causes demotivation. What is particularly offensive to the craftsman is discrimination based on his level in the organizational hierarchy; discrimination that basically says that others may have something that he cannot because they are better or more important than he is.

The men's room in a project field office was once the cause of a major loss of motivation in the field work force. Craftsmen who had come or been sent to the office to interface with engineering and procurement staff employees regarding technical and delivery information naturally used the office men's room when necessary. A middle management employee, apparently reacting to what he felt was an encroachment on one of his privileges, convinced the office manager to prohibit "those field hands with mud all over their boots" from using the restroom. A sign promptly went up on the door reading "THIS RESTROOM FOR OFFICE PERSONNEL ONLY. ALL OTHERS USE FIELD TOILETS." Word quickly spread among the craftsmen and did a great deal of motivational damage. The accompanying loss in productivity was much more costly than assigning a full-time cleaning attendant to the restroom would have been.

On another project a similar sign went up on the office soft drink machine. Some craftsmen were using it to purchase their lunchtime and afternoon break drinks. As a result, the machine was often empty before the day was out, leaving office staff without drinks—depriving them of one of their privileges, some felt. When the project manager learned of the problem, he quickly took action to prevent it from becoming a major demotivator. His investigation found that there were not enough drink machines in the field (the quantity had not been increased with the growth of the work force) and those that were there were not adequately shaded from the sun. Correcting both shortcomings solved the problem.

Granting special privileges, or more importantly, withholding what could be considered a basic need, on the basis of rank is fallacious reasoning. The same is true for variance in rule enforcement according to rank in the hierarchy, as discussed in the previous section.

That rank should not have its privileges does not mean that rank or hierarchical position is unnecessary. On the contrary, rank must be evident because direction must be provided by those responsible for successful accomplishment of the objectives. The project manager must be responsible for the achievement of overall project goals, the craft superintendent for the achievements of those in his craft, and the foreman for the achievements of those in his crew. With rank comes authority; with authority comes responsibility and accountability. The hierarchical structure allows organizations to function efficiently. But granting false status through privileges of rank is divisive and costs more in lost productivity than can ever be saved through the improved performance of the few individuals gaining the status.

THE SATISFACTION VALUE OF PLEASING

Another reason for management to promulgate realistic work rules is because everyone in a civilized society is taught that to gain acceptance, one must be law-abiding. People want to be accepted and so they obey the rules. Most people also wish to please those in authority: their parents when they are children, their teachers when they are adolescents, and their supervisors when they are adults. This is because they were taught that pleasing and being accepted are the routes to reward, even if the "reward" is sometimes only the avoidance of discipline or criticism by their peers or superiors.

In his environment, the construction project, the craftsman also wishes to please and to be accepted. He expects that if he obeys the rules, he will achieve both goals and have a continuing opportunity to practice his craft. In a fundamental sense, he is actually motivated by the act of obeying the rules. This elementary accomplishment gives him a measure of ego satisfaction. He also, depending on the climate set by management on the project or in the company, may expect that his actions and behavior, coupled with his knowledge and skills, will gain him an eventual promotion.

EVEN PRAISE AND DISCIPLINE

Once again, the manager can create a positive situation simply by avoiding the negatives. In addition to remaining alert for instances of uneven application of the rules, he should be watching for opportunities to give praise and recognition for abiding by the rules. The demotivating effect on craftsmen when managers do not take corrective action can cause insurmountable problems with the entire work force. When infractions of the rules occur, discipline should follow promptly, and discipline should be consistent throughout the organization. The treatment of an office worker who takes a calculator should be no less severe than that of a craftsman who takes an electric drill, for instance.

Also, if one employee appears to be "getting away with" something and management tolerates it, the craftsmen will see it as preferential treatment of the offending individual. The result is simple and irreversible. The craftsmen will simply lose their trust and confidence in management. They may also conclude that management is weak and lacks the backbone to assure compliance with the rules. All this, in turn, will have a demotivating effect, with the craftsmen wondering what things will be like when the going gets tough, when materials, tools, and equipment must be carefully coordinated and when people must be relied upon to do what they committed themselves to do.

Preferential treatment of one employee also causes the rest of the organization's members to feel less needed. The feeling of belonging diminishes, and, eventually, the craftsman says, "They've given him preferential treatment that they've withheld from me. They care about him but not about me, so why should I care about them?" Motivation shrinks further, and productivity suffers.

The need for even treatment for everyone cannot be overemphasized in dealing with today's work force. The craftsman believes that he is as good as anyone else he works with and that he deserves the same treatment, whether it be reward or discipline.

And when his supervisor, who represents authority, responds with praise for something good the craftsman has done, or criticism for something bad, the craftsman's belief that he is being given equal treatment is validated. He knows how the manager has recognized good work by other members of the organization. Perhaps they have been praised in the presence of their peers for having achieved some special goal. He expects to be similarly treated with regard to his own special accomplishments. When he does achieve, the reaction of the manager reveals to him whether he is considered an equal member of the group to which he perceives himself to belong. The act either reinforces or threatens his feelings of belongingness. If the treatment he receives is consistent with the treatment accorded other members of the group, his belief in himself and in management is reinforced. If he is treated differently, then his belief diminishes. In the latter case, he comes to one of two conclusions: (1) he made a mistake and he doesn't belong, or (2) management was wrong in not recognizing him as a bona fide member of his group.

The end result in both cases is demotivation. The craftsman feels that if management doesn't think enough about him to recognize his accomplishments, then why should he care about doing a good job? This is another reprise of "If they don't care about me, why should I care about them?"

TELEVISION AND THE SELF-IMAGE

Television is, to a large extent, responsible for the positive image the craftsman has of himself. While he is at home, relaxed and in a suggestive frame of mind, the craftsman watches television actors playing roles in situation comedies, dramas, and commercials. These actors are portrayed in many of the same situations he encounters himself. And he sees them in middle-class and upper-middle-class environments, not in shabby clothing or eating poorly but dressed neatly and eating well, driving late-model cars, and going out to nice places.

What is more, he puts credence in the scenes and skits he sees. The actors look just like him. He pictures himself in the same situations. They become goals for him to strive for, and, sooner or later, many of them become real to him. Life begins to imitate art. Because what he sees on television gives him cause to believe he is just as good as anybody else, can enjoy life the way it is depicted on television, he perceives himself that way and indeed becomes it. He then expects, and in fact demands, that he be treated equally with everyone. Although this perception has been conceived in the personal side of his life, it quickly spreads to his work life also. On the construction project, the craftsman looks, therefore, for the rules under which he works to be applied equally and consistently to all others on the job.

POLICIES AND PROCEDURES

To the craftsman, consistent management means consistent policies and procedures. Probably the most important policy to him pertains to what, when, and how the company is going to pay him. When he hires on, the craftsman is agreeing to exercise his skills diligently at a certain workplace for a certain number of hours per day and a certain number of days per week. He is also taking on certain other obligations with respect to his employment, such as getting to work on time. This is a simple requirement but frequently one that requires a lot of planning and coordination because of car pooling and distances traveled. In many instances, the craftsman must get up as early as 4:30 A.M.

so he can drive fifty or sixty miles to the job site and be certain of arriving on or ahead of time. He accepts this as part of the deal, one of his responsibilities.

In turn, management has a number of obligations to the craftsman, and one of them is to pay him for the number of hours worked at the agreed-to wage. This constitutes the other half of the agreement between management and craftsman.

Craftsmen place a high priority on upholding their end of the bargain and expect management to do the same. For them to act otherwise, they know, would constitute breaking the deal and could invite disciplinary action or discharge. When management apparently fails to put the same high priority on its end of the bargain by, for example, consistently failing to distribute paychecks on time or, because of deficient clerical procedures, making frequent errors in totaling the hours worked or the wage rate to be paid, craftsmen are disappointed. They figure, quite justifiably, that if they can get to work each day on time, then management can prepare a correct paycheck on time. When they do not, craftsmen conclude that management simply doesn't care whether they get paid properly or, worse, is incompetent.

The ultimate thought in the craftsman's mind is, again, "If they don't care about me, why should I care about them?" This expression illustrates the heart of the craftsman's desire with regard to his employment: he wants to be needed, to belong, and to have his employer care about him as a person. He does not want to be a robotlike presence that simply performs assigned tasks.

The good craftsman, exposed to continuing poor treatment, will leave such a project as soon as he can locate another that is managed better. In the meantime, however, his efforts are characterized by the above attitude and the demotivation that results from it.

PERSONAL AMENITIES

A construction project organization brings together a multitude of diverse talents for a finite period of time to accomplish a single overall objective. Some of the work performed by the organization's members is below the ground, some above it; some is outside, some inside. Some rules, such as those for safety, are different for different situations; obviously, requiring a craftsman to wear a safety belt while working 30 feet up in an open structure is a sensible rule, but requiring his helper working at ground level to wear one is not. Other rules, such as those governing the personal amenities, should be the same for all members of the organization regardless of their function or work location. If the manager allows the unrestricted drinking of coffee in the office, he should allow it in the field also. If the practice is abused in either place, he should take corrective action. If he sets a specific time for the ten-minute coffee break in the field, he should set the same time for it in the office. If he's not going to allow coffee-making machines in the field shanties, then he shouldn't allow them in the office, either.

PERSONNEL RULES

The craftsman believes that personnel rules are made for a purpose related to efficiency of operation and, therefore, should be complied with by everyone. The manager cannot grant his secretary an authorized absence for, say, a car breakdown, if he does not accept a similar occurrence as a valid reason for a craftsman being absent.

The craftsman does not expect or want to be treated any better than anyone else, but he does want to be treated equally. In his mind, he is just as important as anyone else on the project and being treated the same confirms this belief. If he is treated less than equally, he begins to have thoughts such as, "They don't recognize me as equal to others," "I'm not needed as much as others," and "I'm not as good as others." Any or all of these feelings demotivate him and cause him to produce less.

Some rules are procedural in nature and grow naturally out of personnel policy. Another responsibility of management, one that is frequently overlooked, is to explain these rules to the craftsmen so that they are clearly understood.

Suppose, for example, that management has a policy that states, "Excessive unauthorized absenteeism will be cause for discharge." As part of this policy, craftsmen should be told that if they have a good, valid reason for being absent, they should telephone the job timekeeper. The timekeeper can then advise the craftsman's foreman, and arrangements can be made to work around the resulting manpower shortage.

If such a procedure is not in place and the craftsman has not been instructed to call in, the foreman may conclude that the craftsman has decided to quit and take steps to replace him. This will create a problem for the absent craftsman when he returns. It will also create a problem for management because there is always a loss of productivity when a new employee is hired.

CONSISTENT STANDARDS OF ACCEPTANCE

Still another responsibility of project management is the development of consistent standards of acceptance with regard to personal conduct, safety awareness, and workmanship.

Forward-thinking management publishes standards of personal conduct and goes over them with each new craftsman when he is hired. In many organizations, the new employee is given a thorough audiovisual or verbal orientation and then given a pocket-size handbook to keep and review as necessary. The key point here is that after such an orientation, there should be no doubt about management's policy concerning pay periods, for example. And there should also be no doubts about what is expected of the craftsman. Standard material in the handbook might include job rules regarding smoking, absences, tardiness, apparel, safety gear, drugs, alcohol, or weapons on the job. It should also spell out the procedures to be followed as part of a gripe-handling policy. Discipline to be expected for breaking the rules should also be clearly stated.

If, for instance, possession of drugs or intoxicants on the job is prohibited (and, of course, it should be), then the craftsman should be told this before he is hired. If such possession is a termination-level offense, he should be told this, also. Then the craftsman knows that if he is found with beer on the job, he will be out of a job. Most craftsmen know that the use of drugs and intoxicants on the job reduces proficiency, induces recklessness, reduces alertness, and, ultimately, causes accidents. And a craftsman who is a hazard to himself is also a threat to those around him. Most craftsmen, therefore, welcome rules of this sort as indicative of a well-run job and of management's concern for the safety of the work force. Strict rules in this area are viewed by craftsmen as a benefit to all. Every industry has a certain number of troublesome employees, and construction is no exception. Firm discipline is often required, and dealing with this group is covered in more detail in Chapter 8.

Similarly, if wearing safety glasses at all times is a job safety rule and craftsmen are expected to be aware of this and comply with it, they should be told so during the hiring process so that there will be no question that they were properly instructed. Many companies, especially on projects in more litigious states, confirm the craftsman's receipt of this type of information by getting a signed acknowledgment during the orientation procedure.

In each of these areas—personal conduct and safety awareness—it is vital that the standards of acceptance be clearly delineated to the craftsman and that his acknowledgment that he understands them be obtained at the very beginning. If this is done, then each craftsman will know that he is subject to the same rules and standards of acceptance as everyone else. He will also know that he is subject to the same discipline for not meeting them as everyone else.

"SINGLING OUT"

As before, management's continuing responsibility is to assure that rules and discipline are consistent from one employee to another, from one department to another, and from one level in the hierarchy to another. Perceptions of the craftsmen are critical in this aspect, also, and the manager must cautiously avoid even the perceptions of "singling out" to develop. Obviously, if a craftsman is singled out, or even if he *feels* he is being singled out and treated unfairly—especially if it is for personal or political reasons—it is, at best, a demotivating experience for him and any fellow craftsmen who are cognizant of what is going on. At worst, it could provoke an aggressive response, even an attempt to "get even" by acts of sabotage—defacing finished work, cutting hoses, or putting sand in fuel tanks. It could even lead to open unrest and an organizing attempt by a union.

BUILDING A CASE

Management should also avoid any actions that might be perceived as "building a case against" or "trying to get the goods on" an individual or a group. Just as letting someone get away with something is instantly recognized by craftsmen, trying to get "at" someone by fabrication, entrapment, or "setting him up" will also be recognized and will do more harm than good. Such action goes against the craftsman's sense of fair play, a sense which is as prevalent in the construction work force as it is anywhere in American society.

This is not to say that effective investigative action shouldn't be taken to root out troublemakers because the good craftsman wants them off the job as much as, if not more than, management.

ELEMENTARY SCHOOL ATMOSPHERE

Obviously, a busy construction site is not the place to permit a totally unrestricted environment; too much is at stake, especially in regard to accident prevention. In a sense, a construction site is its own society, created anew each time people leave and others are hired. Because of the coordination required between the crafts to progress the work, procedures for accident prevention, security, and other standards must be set. Care must be taken, however, to maintain an awareness of the maturity level of those whom the rules are intended to govern. Job-site rules should be designed so as to avoid generating an elementary school atmosphere and creating the perception in the minds of the craftsmen that they are being treated like children. No one should have to raise his hand to get permission to go to the bathroom or get a drink of water, and the manager should continually be on the lookout for a proliferation of rules that tend to treat the craftsman like a child.

HUMANE EXCEPTIONS

The world is not perfect, and rules get broken from time to time. In setting policy, therefore, it is important that management not get caught up in its own rules to the extent that it can't make humane exceptions that craftsmen can recognize, understand, and relate to positively.

Consider the project on which a craftsman was absent from the job for three working days without calling in to explain his situation and obtaining an authorized absence. Assume, in this case, that the rules called for him to be terminated and his name removed from the job payroll—a standard operating procedure that everyone hired had been told about. In many instances, the explanation would have been that the craftsman had decided to go to work on another project or go fishing or hunting for a while. In this instance, however, the craftsman had been involved in an accident that had physically prevented him from making a phone call for three days. When he was finally able to call in and explain what had happened, should he have been told he was terminated because of the three-day rule? Of course not.

7

Sources of Frustration

Frustration denotes the feelings of discouragement and ineffectiveness a person experiences when he fails to perform a task or accomplish a goal he has set out for himself. In many people, the stronger the desire to reach a goal, the greater the frustration level when that goal is not reached. Also, a strong desire to accomplish a simple goal or to perform an easy task produces more frustration when that goal is not reached than does a strong desire to reach a complex goal or to perform a difficult task. This is because the expectancy of accomplishing the simple goal is greater than is the expectancy of reaching the more complex goal.

PRIMARY SOURCES

On a construction project, frustration of one kind or another can affect everyone in the organization from top to bottom, but, in many ways, the most intense frustration is experienced by the craftsman. This is because he is the least in control of the factors that cause his frustration.

There are three primary sources of the frustration that lowers the craftman's effectiveness.

1. factors that he can control
2. factors that are beyond anyone's control
3. acts by others

Self-Generated Frustration

The first source of the craftsman's frustration are his own shortcomings and are due solely to his inability to perform a task as well as he would like to. If he is a journeyman carpenter, he takes pride in making accurate measurements with his ruler. He is disappointed when he makes an error because he knows that with better attention, he could have prevented it from happening. He may reprimand himself to assure more concentration next time.

If he is a welder, he may be frustrated when he produces a weld that must be ground out and replaced. As a journeyman, he knows how to make a defect-free weld, but occasionally his concentration slips and the weld fails X-ray examination. If the defective weld was made in a difficult location or a difficult position, he will be mildly frustrated. If it was made under less severe conditions—for example, in a shop at a workbench—he will be very unhappy with himself for the same failure.

In each case, the frustrating element is the dissatisfaction the craftsman experiences when he knows he can do something better than he has. Like a baseball player, he keeps his own personal batting average. He knows what constitutes acceptable performance from him and what will qualify him for slugger status. This source of frustration is under the craftsman's control because he knows that it is within his power to correct or improve his performance. He knows, also, that when he improves his proficiency, his frustration will be replaced by its opposite—satisfaction. Practice, an awareness of what constitutes good performance at his particular skill level, and concentration will minimize this source of frustration in the seasoned craftsman, allowing the ego satisfaction that comes with achievement to prevail.

Factors Beyond Anyone's Control

The second source of the craftsman's frustration is due to influences beyond anyone's control. A rainy day that forces a project to be shut down falls into this category. An craftsman who has, for example, just driven fifty miles to get to work and meets a deluge of rain as he walks in the gate is disappointed, to some degree, because of the drive. More indirectly, he had hoped to work that day, earn his pay, and derive satisfaction from practicing his skill.

Accidents and catastrophes also fall in this category. Work on a pollution-control project was once halted for several weeks when a fire in a fabrication shop destroyed several almost-completed fiberglass scrubber section fabrications. The molds as well as the fabrications had to be reconstructed. Craftsmen at the installation site were understandably disappointed at having to wait, but the frustration level was not high because they recognized that not much could be done about it. In this particular instance, the employer turned the negative experience into a positive one by shipping the remanufactured pieces via special routing and with escorted expediting to shorten the en-route time by several days. The craftsmen recognized this extra effort by management to get the project rolling again, and a higher motivational atmosphere resulted.

In other instances involving delays, craftsmen will not perceive the act to be unavoidable unless management takes the time to explain what went wrong. The manager on a highway bridge project once lost credibility with his work force because he did not tell them why the structural steel beams were late. They thought it was poor procurement management on his part; actually, the delay was from causes beyond the control of anyone on the project staff. Three railroad flatcars with the needed steel beams on them had been partially covered up in a switching yard by drifting snow during a storm. It took two weeks to get them dug out and sent on their way to the project site. The frustration the craftsmen experienced could have been abated, and they would have accepted this as an unavoidable delay, if only they had been informed as to what was going on.

When craftsmen encounter these potential sources of frustration, although they may not like them, they realize that not very much can be done about them and so they accept them philosophically. As a result, the frustration they experience does not have a serious effect on their motivation.

man feel inferior, unskilled, stupid, or, in some other way, incompetent, the craftsman's feelings of self-worth will be jeopardized, with resulting dissatisfaction and frustration. On the other hand, a checker with a professional demeanor, through proper words and actions, can show the craftsman any errors he is making and how to correct them in such a way that the craftsman will feel the checker is helping him improve his performance and avoid mistakes.

SUFFICIENT COMPETENT PERSONNEL

Once proper briefing of both the checker and the craftsman is allowed for, the manager must make sure that sufficient personnel are available to perform the required inspections expeditiously. Nothing is worse than making a craftsman who has reached a holding point or checkpoint in some installation stand idle for several minutes or hours waiting for the checker to arrive. He may not become frustrated in the time it takes the checker to actually perform the inspection, but he will be extremely displeased if he has to wait half a day for the inspector to come by. This illustrates the difference between the unavoidable and the avoidable frustrations described above. With proper orientation, the craftsman understands that the inspection must be performed and accepts that fact. His frustration is minimal. But he knows that with proper coordination and sufficient competent personnel, checking can be performed in a timely manner. When this does not occur, he becomes frustrated and is demotivated, and, eventually, a deliberately unconcerned attitude and loss of productivity result.

MANAGER'S ROLE

The task of the manager, then, is to preclude or remove all sources of frustration within his control. Much of the material presented in earlier chapters about getting the plan, materials, tools, and equipment to the point of the work at the right time and in the correct order is intended to prevent frustration from developing in the first place. All of these sources of frustration are ones that the craftsman knows can be controlled by competent management.

To reiterate, McGregor's Theory Y characterizes the craftsman as ready, willing, and able to work if permitted to do so by management. But when the craftsman sees little being done about the controllable factors, his frustration results in an attitude characterized by "If management doesn't care about getting the job done by bringing all the necessary components to the point of the work, why should I care about getting my job done?" In order to maintain the craftsman's interest in doing his job productively, the manager must set the pace by doing his job productively. This leads to an important concept.

The determination as to whether the craftsman will correspond to Theory X or Theory Y is, indeed, made by management. If management does not do its part in removing the dissatisfiers and sources of frustration, the craftsman will be driven directly to the Theory X attitude. He will not want to come to work, not want to work once he is there, care very little about production, and think mostly about the time remaining to quitting time.

On the positive side, satisfaction, the opposite of frustration, occurs when management has done its job of coordinating the essentials at the point of the work. The craftsman is able to use, without undue interruption and delay, the skills he possesses. He uses them well and enjoys the experience. He is on the project early in the morning, eagerly looking forward to a day of productive effort. He's proud of the end result of his day's work, and, when the project is completed, he looks back with pride at his role in it. He feels good about himself; he is satisfied.

Well-managed projects are usually remembered with pride and satisfaction by the craftsmen who worked on them. When the fiftieth anniversary of the completion of the Golden Gate Bridge in San Francisco was celebrated, among those participating were several craftsmen who had worked on the bridge. In interviews with the press a half century after the bridge was opened, they expressed a lasting pride and satisfaction in their roles in the construction of that civil engineering landmark.

PERSONAL RELATIONS FRUSTRATIONS

Besides poor coordination and work flow, there are other correctable sources of frustration. Although they do not have much to do with the direct events that produce the completed project, they do affect the craftsman's attitude and motivation. They have their roots in supervisory styles that are oblivious to or uncaring about the human side of management. These styles affect the craftsman's ego and bring personal issues into conflict. Unlike problems in logistics which can be solved easily by the manager, these frustrations, once created, take much longer to deal with effectively; and sometimes a cure is never achieved.

A closer look at some of the actions or inactions by management that can really frustrate and thus demotivate the craftsman will provide greater insight into this aspect of the problem and help assign dimensions to its importance. It is worth noting here that many times, the manager is unaware of the negative effect some of these actions or inactions can have on craftsmen because he has not taken the trouble to empathize.

Consider the paycheck problem discussed in the previous chapter. Everyone makes an occasional mistake, and the craftsman is forgiving when his paycheck is a few hours short once in a while. After all, he understands that a busy foreman can forget a half hour of overtime which, on a project site without a time clock or brass system to serve as a double check, can go all the way through the timekeeping procedure without being discovered. The normal remedy is for him to point it out to the foreman who, unless the error is major, then has the shorted time paid in the next check. But if the same thing happens more than once or twice, it indicates that something is wrong procedurally, something for which the craftsman holds management responsible, something he knows management can correct.

When corrective action is not taken, the craftsman's frustration level rises because he begins to take it personally. His feeling is that he has earned his pay, and, in effect, someone else is holding it. Moreover, he may be in a situation in which he has made commitments based on what he earned and expected to be paid for during that pay period and doesn't want to suffer the embarrassment of breaking these commitments.

Consider, also, the instance in which job rules require that the craftsman be at his workstation until five minutes before quitting time. The last five minutes are to allow him to put his tools away, clean up, and head for the exit gate. Imagine his chagrin when he gets to the gate at quitting time and finds foremen, general foremen, and sometimes even superintendents already in their vehicles, heading toward the main road, getting an early start on the rush. He rightly feels that the rules should apply equally to everybody (consistent management), yet while he was still working, the bosses were on the way out the gate. He feels exploited and discriminated against—after all, he's as good as the rest of them, so why should they take advantage of their position to break the rules and leave early. He wants to get home to a shower just as much as they do—perhaps more.

Or consider the craftsman on a job with a safety program that requires everyone to wear hard-toed safety boots for foot protection and safety glasses to protect against flying debris. Neither piece of gear is particularly comfortable—especially on a hot, humid day—but he believes management put the rules into effect because they were necessary to prevent accidents.

Imagine his reaction when he sees some important visitor, perhaps the "big boss" from the home office, making a tour of the project wearing designer sunglasses and imported leather loafers. The craftsman thinks, "Isn't this VIP just as exposed as I am to getting dust and debris blown against his fragile glasses when he's out on the job? Isn't he just as susceptible to stubbing his toe or having something fall on his foot?" The answer is yes. Then why isn't the VIP wearing the same safety protection that the rest of the work force has to wear? The answer is obvious: because he is important, because he is above the rules even though the rules were made for safety purposes (at least that is what the craftsman has been told). The craftsman's frustration level rises.

These are examples of sources of frustration that the manager can eliminate if he takes the time to listen to, to get to know, and to talk to his craftsmen. Even more easily, he can see it all for himself if he will simply observe with an empathetic eye; and then take positive action to remove the sources of the frustrations. When he does this, he will see a better than proportional response to the actions he has taken.

To be successful in eliminating sources of frustration due to inconsistent application of rules and standards as in the above illustrations, the manager need only develop the capacity to think in terms of the effect his actions will have on craftsmen, see it from their perspective. In developing his rules, procedures, and standards of acceptance and, perhaps more important, conveying them to craftspeople, the manager need only ask himself questions such as, "Am I willing to have them applied to me along with everybody else?"

POOR MAINTENANCE OF TOOLS AND EQUIPMENT

Other management oversights, oversights that relate more directly to the day-to-day operations of the project, cause not only frustration but direct loss of productivity through forced idle time. Remember that most craftsmen are happiest when they are exercising their skills, and they're unhappy when they are delayed or prevented from doing this.

If a craftsman withdraws an electric drill from the toolshed, for example, and finds, when he tries to start it back at his workstation, that it doesn't work, he's not only frustrated but he's also lost the time it takes to return the drill to the toolshed and get another. If he's performing a vital operation in the sequence of a crew operation, the whole crew is delayed for the same amount of time. Again, if it is an isolated occurrence, he accepts it as the exception rather than the rule. When it happens frequently, he knows that management is not doing its job and concludes that the manager does not regard the need for a smooth, efficient operation very highly. What follows is a decline in his enthusiasm for his work and the project and his interest in doing a good job for his employer. If they don't care, why should he? The result is a drop in his productivity that goes far beyond the fifteen or so minutes it took for him to go back and find a drill that worked.

POOR COORDINATION

Consider the disappointment carpenters experience when they come out in the morning to finish up a wall form so concrete can be poured the next day and they find rodbusters still tying reinforcing steel in place. If it takes a full day to finish the form, then one of two things will happen: (1) they will have to stay over to get it ready for the scheduled pour or (2) the schedule will be delayed.

In either case, the time lost is only the most observable result. A six-man crew with nothing or, at best, only *busywork* (activity designed chiefly to make the crew look busy without actually producing much or any finished work) to do is costly. Every ten minutes, the crew wastes a full man-hour. Far more costly, however, is the demotivation caused by the frustration. Demotivation is easy to ignore in its early stages, and a poor manager will ignore it because it is difficult for him to acknowledge that he is the cause of it. He hopes something will happen to make the problem go away or that someone else will find a solution and save him from having to determine which supervisors are not doing their jobs. By the time the manager realizes that none of the above is going to happen and he faces up to taking the necessary actions himself, it is frequently too late. He has either created a lethargic atmosphere that will prevail as long as he is in charge, or the good craftsmen will already have recognized the poor quality of the job supervision and fled to a better-run project. Each construction project is immersed in its own unique motivational atmosphere. This atmosphere, like the one that surrounds our planet, is fragile and easy to damage. Once damaged, it also is difficult, if not impossible, to restore to its original healthy condition.

How much miscoordination or lack of support of the craftsman is too much? How many repetitions of an error in judgment is too many? The manager must decide this for himself. If he is sincere and truly dedicated, he must ask himself honestly how many times a week or month he would tolerate a particular correctable situation before becoming frustrated himself. He must empathize. The level at which he decides corrective action must be taken will determine the atmosphere on the project, and that atmosphere will determine the success of the manager, not only on this particular project, but probably in his entire career.

REMOVAL OF SOURCES OF FRUSTRATION

The removal of sources of frustration creates a phenomenon that is readily observable. When management removes sources of frustration, craftsmen do a better job. Efficiency improves, productivity improves, the craftsman's sense of accomplishment improves, and his ego satisfaction soars.

Increase in ego satisfaction causes craftsmen to look for ways to improve the flow of their work. They will make their part of the operation more efficient so they can attain an even greater sense of satisfaction. It's no different than the winning football or baseball team. The members get an immense amount of personal ego satisfaction out of winning. It makes them feel so good, they try even harder to win the next game. Setting goals and deriving ego satisfaction from reaching these goals—that's the fundamental driver of motivation, whether on the ball field or in the construction field.

On the construction project, removal of sources of frustration creates a win-win situation. The craftsman wins because he obtains more ego satisfaction at more frequent intervals. Management wins because the work flows more smoothly and efficiently, more units of work are produced per time interval, and overall higher productivity levels result.

8

Fine Tuning the Work Force

Organizations and individuals tend to amalgamate because of a mutual expectation that the individual will benefit the organization and the organization will benefit the individual.

MUTUAL EXPECTATION

In the construction industry, the contractor needs the skilled craftsman in order to build a project. He therefore offers employment at a wage that will attract craftsmen. In return, he has a right to expect the craftsman to be productive and qualified in his skills. The craftsman, on the other hand, possesses the requisite skills and expects to practice those skills for the benefit of the employer for the wages promised.

There are, however, departures from expectations on both sides. There are abusive and illegal practices by some employers, and there are misfits, nonperformers, and troublemakers among craftsmen.

EMPLOYER'S DILEMMA

When the employer errs, the employee can obtain zealous assistance very quickly from the state labor board, from any one of several supportive federal agencies, or from a union if he is so represented. But the employer has a dilemma. There is no functionally effective corresponding remedy available to the employer—he is essentially on his own. Not only must he find ways to identify the few craftsmen who, for one reason or another, cannot or will not hold up their end of the bargain, but he must also avoid any appearance, whether founded or not, of noncompliance with labor-related laws, rules, and regulations.

The only clear fact in an otherwise murky area is that the manager, when faced with a dilemma involving employee misbehavior, poor work habits, or lack of skills, must take positive action in order to avoid demotivation of the rest of the work force and a resulting loss of productivity. As mentioned in earlier chapters, the great majority of craftsmen want to do their jobs and exercise their skills without hindrance or disruption. They become frustrated and demotivated when prevented from doing so by outside influences, including other craftsmen who cause trouble or get in their way.

The manager who ignores the imperative presented by such a situation and fails to act suffers an unrecoverable loss of credibility with the supervisory cadre as well as the craftsmen in everything he subsequently tries to do on the project. He becomes essentially powerless to manage the job from that point on.

Problem employees fall into three basic categories: (1) incompetents, (2) those with poor work habits, and (3) troublemakers.

Incompetents

Incompetents can be found on all levels of the construction industry—from entry-level laborer to project manager. Dealing with an incompetent project manager must be left to someone higher up in the off-project organization to which he belongs and is beyond the scope of this book, but dealing with incompetence within the project organization itself is the responsibility of each level of supervision and is ultimately the project manager's responsibility.

There are two clues to the presence of an incompetent person in a crew or department: (1) a high level of errors is detected and (2) schedules and deadlines are missed. Either one, or both working together, produce still another characteristic of a project in trouble: higher costs. All three symptoms can be analyzed, but investigation of the first two—errors and missed deadlines—will lead more quickly to a direct solution of the problem than will investigation of higher costs. The investigation usually leads to employees who either do not possess adequate skills (because of a lack of aptitude, training, or experience) or who have some skills but not ones applicable to the assigned task. Sometimes, also, it leads to employees who, because of some physical limitation, are unable to perform the assigned work.

The Underqualified

Occasionally, job applicants try to pass themselves off as more qualified than they are because they want the job being offered. In some cases, applicants are sent by a dispatcher from a central hiring reference, such as a union hall or state employment office. The objective of the dispatcher is to find a particular person employment—perhaps because he has been "sitting on the bench," waiting for employment, the longest—rather than to find the most skilled person available for the employer. And sometimes even the skilled people available have the wrong skills for the operations at hand.

For example, a pipe welder who produces a high percentage of flawed welds may simply need additional training because his previous welding skill was limited to flat surfaces, such as structural steel. Or maybe his lack of skill is such that he shouldn't really have sought work as a pipe welder at all.

A carpenter who continually cuts lumber too short (or too long) may actually be at the skill level of a carpenter's helper and need additional training to achieve journeyman status. A bookkeeper who confuses debits with credits may really be qualified as a timekeeping clerk instead. A secretary who takes all day to produce a two-page letter on a personal computer may be a whiz with a typewriter but not have had training in the software being used on the computer, or she may be a computer-illiterate.

All these misfits have been discovered in the workplace at one time or another. Once, on an airport terminal project in New York, column anchor bolts on three separate foundations had to be jackhammered out because the lead carpenter, in charge of locating the bolts in the concrete forms accurately, confused a dimension on the plans. He continually read (10") as (1'-0") and had apparently been doing this for quite some time because it was discovered that he had very little knowledge of blueprint reading.

The Physically Unable

Another category of incompetent employees are those who are unable to perform the work physically for one reason or another.

In a few crafts, age and experience are definite advantages. Some contractors even set up special "master" or "lead" designations and pay grades for this group. For example, a millwright with several years' experience aligning large, high-speed machinery such as that found in the pulp and paper industry is eagerly sought after by managers. The older he is, the more valuable he is; many of the best are in their sixties and, in some cases, seventies. In these cases, physical strength and endurance are not as important as the craftsman's skills, judgment, and prior experience. On the other hand, a carpenter handling heavy timber, or an ironworker in an erection crew, must have the physical capacity to do this work. Through conditioning and careful stewardship of their bodies, these craftsmen can function productively well into middle age. Some, however, cannot. The latter should be identified and assigned to less physically demanding work, if available, before they become injured or disabled.

On a large project, a physically limited carpenter can be assigned layout, measurement, saw shed, or blade-sharpening duties. It should be emphasized that these are not make-work activities but services that are fully required and that require the skills of the craft. Many contractors put their most trusted craftsmen in the saw shed, for example. The *saw shed* is the place where repetitive operations and production cutting of lumber and timber are performed; large quantities of a particular size or length piece are cut and fabricated at the same time. A mistake in cutting, say, 100 pieces of form stud defeats completely the advantages of making such a production run.

Some job-site tasks, however, simply require strength and endurance, and those who cannot perform should be laid off if no other work can be found for their skills. This group of people probably should not have solicited employment in the particular craft to start with and certainly should not have been hired without verification of their recent work experience.

Also part of this category of incompetents are those who, because of an injury, cannot perform the work of their craft. This group is hard to detect before hiring because they usually do not admit on their employment applications that they are not physically fit. Their goal is to get on the payroll and hope that no work exposing their limitations will be assigned to them.

In some cases, also, a person may deliberately conceal a prior injury with the intent to defraud the employer. His goal is underhanded, and, once found out, he is scorned by craftsmen who are honest and committed to achieving an accident-free safety record. His plan is to collect pay without working. Soon after being hired, he places himself in some plausible activity or setting, feigns an accident, and claims he has been injured. This ploy, if successful, enables him to collect workmen's compensation and disability benefits from the employer's insurance or from state funds. Most of these cases involve claims of back injuries, which are difficult to diagnose accurately.

One contractor, tired of being taken advantage of in this manner, subjected every previously unknown applicant to a pre-employment medical screening by an industrial nurse. This was not an actual physical examination but, rather, a compilation of the applicant's medical history and a visual study by the nurse's experienced eye. In addition to inquiring about obvious features such as scars and missing fingers that would reveal prior accident history, the nurse conducted one important test.

During the screening, the applicant was seated at the side of the nurse's desk while his medical history was being taken down. Midway through the interview, the nurse would "accidently" let her pen fall off the desk to the floor. Sometimes she even got it to roll under the desk. Those who didn't volunteer were casually asked to pick the pen up for her. If the applicant seemed to have trouble moving and bending down to pick the pen up, she would ask a series of more detailed questions and have the applicant perform a few simple exercises. If back problems were confirmed, the applicant was required to undergo a physician's examination and provide a fitness report before being considered further for employment.

Poor Work Habits

Craftsmen with poor work habits are usually easy to spot. Among them are the "walkers," the "talkers," the "lazies," and the "layouts."

"Walkers"

The walker is one who spends most of his time during the day walking around the job rather than working. He can be found taking the long way to and from the toolroom when he is sent to pick up supplies, using toilets and drinking water on the opposite side of the job, making frequent trips to the first aid station for aspirin or to the timekeeper to ask questions about his paycheck, sometimes just making a circuit of the job work areas. He walks at a fairly rapid pace to give the appearance of being in a hurry because he believes this creates a good impression with the supervisor (and, to the unwary supervisor, it does). He always carries a tool or a piece of material as a prop to present a productive appearance and, to the casual observer, seems to be quite busily engaged. At the end of the day, however, he hasn't accomplished anything.

"Talkers"

The talker, on the other hand, usually remains at his work station but readily stops whatever he is doing to talk—with his fellow crewmen, with passersby, with supervisors—with anyone who will stop to chat. A simple "Good morning!" can stretch out to fifteen minutes. In some instances, the talker simply lacks self-discipline. In other instances, he is deliberately trying to avoid performing his assigned work. Sometimes the walker and the talker get together. In general, however, the talker is more damaging to productivity because he effectively stops the person with whom he is conversing from working also.

"Lazies"

The lazy worker usually reveals himself quickly. If he is shoveling dirt, he'll move a shovelful (or half a shovelful) and then stop to look around for a minute or so (longer if he doesn't see anyone watching him). If he is sent to pick up materials, he walks more or less straight toward the toolroom but at the slowest forward speed possible, in sharp contrast to the walker. He, too, makes frequent trips to the toilet and to the water barrels, all in slow motion. Occasionally he finds a place to hide out for a few minutes or a few hours, depending on the proximity of his supervisor. Once, on an office building project, a few craftsmen spreading gravel on the roof found that the inside of one of the adjacent large air-handling units was a good place to hide, so they took turns doing it until an alert supervisor noticed his gravel-placement costs were running 20 percent higher than usual.

Layouts

The layout is easily detected because of his attendance record. Of course, unless fraud is involved, he does not get paid for the time he misses. However, he does have a negative impact on production. For instance, assuming he is an essential part of his crew, he must be replaced by someone else on the days when he "lays out" (does not come to work). Just as the football team does its best when all the first team is able to play together, so the construction crew is most productive on days when its regular crew members work together. In addition, no matter how short his absence has been, when the layout comes back to work, he is not immediately effective because he must become reoriented to what has taken place while he was away.

On one petrochemical project in an area of the country in which large families lived together—and worked together—the project manager was puzzled by the absentee patterns being experienced on the job. Attendance was good on Mondays, Tuesdays, and Wednesdays but very poor on Thursdays and Fridays. The pattern seemed to be repeated every week, and he believed it was causing losses in productivity because, at the same time, project progress was falling behind schedule and costs were skyrocketing. Carefully studying the attendance records, he discovered that there were several instances in which employees with the same last names were absent during the last two days of the week just about every week. The problem was centered in several groups of families. Many of them had as many as five members on the payroll, all in different crafts and crews; and all of them were absent each Thursday and Friday. Interviewing a few of the senior family members, the manager finally discovered what the cause of his dilemma was.

All members of the family typically lived together—perhaps in different houses but on the same property or in close proximity. The older men and women, and the children stayed home to clean, cook, and keep house while the working-age members, both male and female, worked on the construction project. They found that with all of them on the job payroll, by Wednesday they had collectively made enough in wages to cover their living costs, and so they simply took the rest of the week off to fish, hunt, sew, or whatever suited them more than working.

They had no concept of what their peculiar work ethic was doing to the project; they had not even thought about it. In their part of the country, attitudes were relaxed and living was easy. For generations, they had been working no more than necessary to "put groceries on the table." Three days of work a week did this for them. They did not understand, until the manager had a heart-to-heart talk with each senior family member, the negative impact their casual attitude toward work was having. Once they knew, however, they realized the importance of their contribution, and the problem was solved. Actually, it was more than solved, for when they became aware of their impact as a family, they took pride in their attendance record—coming to work began to mean more to them than just working and collecting pay.

Troublemakers

Troublemakers are present, to some degree, in all walks of life. In construction, the problem is relatively minor because the industry has a tradition of self-policing—the craftsmen themselves take care of troublemakers in their own ranks. They do not want the burden of constantly looking over their shoulders to see if their tools are being stolen, their work is being sabotaged or vandalized, their personal safety is being jeopardized, or their welfare is being threatened. They quietly take action themselves, and most times the manager is not even aware that there was a problem.

Sometimes, though, they do not or cannot handle the problem themselves because of the particular circumstances involved, and it is therefore important that the manager recognize the symptoms and position himself to take swift action when it becomes necessary.

IDENTIFYING TROUBLEMAKERS

Troublemakers come in all forms and shapes. Some are loud mouths and readily give themselves away through constant complaining and boisterous behavior. Some are just the opposite —introverted loner types. Some are physically intimidating, bully types, while others are small wiry types with knives or pistols in their pockets. They may be drinkers or dope-heads. They may be psychologically troubled, hate themselves or the world in general, or both. They may simply be unhappy and want everyone else to be unhappy, too. They may be insecure and need the bolstering that a sense of superiority brings. They may be thieves.

They may be craftsmen or supervisors. They may also be staff employees, such as warehousemen or security guards. And they may be female as well as male.

The manager's first indication that there is a troublemaker in the ranks is usually an incident involving damage to work already in place, the theft of personal property or job materials, an injury to a craftsman, or a physical outburst on the project site or in the office.

UNSETTLING INFLUENCE

The common thread in all these cases is an unsettling of the work force and a disruption of the normal work routine. A foreman supervising a thirteen-man railroad-track-laying gang once encountered an employee who refused to perform the work the rest of the crew was assigned, placing heavy timber ties and rails. He simply would not carry his share of the load, and the other craftsmen were upset at the increased burden it placed on them (putting an 8-foot-long cross tie in place is physically demanding for four strong men, much more so when only three are doing the work). Unfortunately, this troublemaker was also the *shop steward,* the union representative on the job, who, by agreement, was entitled to spend a "nominal" amount of time (theoretically about half an hour) each week taking care of union business related to the job. The steward insisted that doing the heavy work tired him to the extent that he could not discharge his union responsibilities, all much to the chagrin of his fellow craftsmen. They were, however, powerless to do anything in this instance because of the fear of reprisal: the steward determined who would work on the job and, more importantly, who would not.

Isolate the Problem

The manager felt that if he called for replacement of the steward by the union, he might still have the problem, just in the form of a different person. Instead, he struck an agreement with the union business agent. The steward, while not being required to perform the heavy work, would, nevertheless, be required to perform a full day's work every day, except for the half hour per week allowed for union business. As soon as the agreement was reached, the manager gave the steward a brush and a bucket of paint and sent him to the far end of the project where the *bumper* (the structure that stops railroad cars from running off the end of the track) was located. His only assignment was to paint the bumper. He was thus effectively isolated for several weeks (he didn't paint well, either), and, when he finished, the manager assigned him another equally remote task. The steward's place in the track gang was filled by a replacement, and the rest of the crew, happy to have a fully participating team member, settled back into a productive routine.

Troublemakers in the Office

Troublemakers in the office can cause as much disruption to the normal project routine as troublemakers in the field and can sometimes do a lot more damage in terms of cost. A purchasing agent who accepts expensive meals and lavish entertainment, free trips in corporate jets, and the use of resort hideaways from would-be vendors and subcontractors compromises his judgment, usually at the expense of the project. He also has a demotivating effect on the remainder of the office staff and even the field work force. Most people realize that the vendors and subcontractors recover the costs of these perquisites somehow through extra billings of some sort, and they question why management emphasizes productivity goals on the one hand and allows these excesses on the other. Why should they strive, they wonder, to save dollars by working more efficiently in the field if, in effect, what they save is going to be spent entertaining members of the office staff?

In addition, many personality types that become involved in this activity cannot wait to tell everyone about their venture, thus causing negative feelings among their co-workers ranging from envy to contempt. In any case, cooperation, coordination, and communication among the staff suffers, with resulting losses in motivation and productivity.

Diminished Credibility

The longer a troublemaker is tolerated on a project, the more management's credibility and ability to manage effectively are diminished. A serious situation left unattended will quickly cause the work force to view the manager as weak and unable to lead. The good craftsmen will leave for other work. The work will, of necessity, be completed by lower-quality and less-able craftsmen, and productivity will fall.

As indicated above, the troublemaker can be found at any level. The general superintendent on a power plant project in California once outfitted the small job-site maintenance shop with enough equipment to rebuild engines, which he promptly started doing—not the engines of the construction machinery but engines from his personal autos and the autos of several of his relatives. What was even more distasteful to the craftsmen was that he brazenly put both his father and his son on the payroll to do the repair work. Subordinate supervisors and craftsmen with honest work ethics and higher standards of conduct quickly lost confidence in the superintendent when they saw what was happening, and it wasn't long before the project manager found out what was going on and discharged him. Job-site promotion of one of the other supervisors immediately turned the situation around and restored the project to its planned cost and schedule track.

DEALING WITH THE PROBLEM EMPLOYEE

When a supervisor encounters a problem employee on his project, he should first analyze the symptoms to identify the root cause of the employes's problem and to determine if he is salvageable. If his work ethic is reasonably strong and his attitude wholesome, every effort should be made to find a continuing assignment for him. In addition, he should be given the opportunity to learn or improve his skills through training. The flat welder can be trained in pipe welding; the home-building carpenter can be taught how to fabricate concrete forms; the computer-illiterate can be shown how to make basic data entry. Anyone already on the payroll is, to some extent, known and "calibrated," as compared to someone new about whom little or nothing is known. (Each time a replacement is hired, learning-curve penalties take their toll on the replacement's initial effectiveness.) On the other hand, if the trouble is traced to dishonesty, an incompatible attitude, or recalcitrance, the employee should be weeded out.

Dealing with the problem employee takes three basic forms: (1) preventive action, (2) corrective action, and (3) adaptive action. Any of these actions can be immediate, interim, or permanent solutions.

Preventive Action

SCREENING

It is always better to avoid a problem than to have to deal with it later. Most contractors today have sophisticated data bases that can display a comprehensive personal and work history on all present and former employees. Usually, previous employment for a period of three to five years is kept up to date. A part of the data base is the *termination code* for each engagement;

i.e., the reason for severance of employment. Code A might be assigned to those who were laid off as part of normal job staff-down; Code B might be for those who quit with appropriate notice. Additional codes are assigned for other categories of termination, ranging from voluntary to involuntary. Before hiring an applicant, his history should be checked; if his prior service was satisfactory, the probability is that the current engagement will be, too. Conversely, if his prior record is clouded, the supervisor can avoid any new potential problem with that applicant by hiring someone else.

In a union situation, prescreening through data-base checking is usually the best technique. If that is not possible, however, the next-best way is to make the business agent aware of the hiring criteria and to solicit his cooperation in referring only craftsmen who meet these criteria.

Some contractors staff a new project by offering reemployment to previously satisfactory employees through mailings and phone calls. Others make use of word-of-mouth recruitment. If the initial cadre is made up of good craftsmen, it is likely that the craftsmen they recommend will also be good.

Contractors who are in a repetitive business, such as making all boiler repairs in a particular operating area, have another system. Over a period of time, they hire and terminate craftsmen until they develop the organization that meets their needs most efficiently. This is called *culling out*. Then they hold on to these craftsmen at all costs, including keeping them on the payroll even when work loads are down.

Absent the above screening opportunities, the supervisor must rely heavily on the information included in the craftsman's job application. (A competent labor relations attorney will provide advice in regard to what information can be required on an application and, more importantly, what data cannot be solicited without risking punitive action through various government agencies.) Careful study of the application, combined with an interview regarding previous employment, coupled with one or two detailed craft knowledge questions, will do a lot to eliminate problem employees.

In uncertain cases, a quick look by the supervisor at the contents of the craftsman's toolbox will usually reveal his skill level. For example, if a carpenter's toolbox contains two different weight hammers, a block plane as well as a jack plane, a ripping saw as well as a crosscut saw, a 50-foot tape and a scribing compass as well as a 6-foot ruler, and they're all used but well-maintained, chances are he is a bona fide journeyman carpenter. On the other hand, an applicant aspiring to hire on in the same category with only a hammer and a nail claw on his tool belt most likely would be better employed as a carpenter's apprentice until he gains the requisite skills and tools. This happens quite frequently, and many contractors take the opportunity to provide craft training to these applicants that will qualify them to become the journeymen they obviously wish to be. It develops into a win-win situation and is one of the more pleasant illustrations of corrective action.

Laying Out the Rules

Those offered employment should be given a comprehensive orientation that includes a review of the job rules. Many employers also present new hires with a set of printed rules and conditions of employment, obtaining written acknowledgment of their awareness and understanding. A list of prohibited activities that normally result in discharge should be clearly and simply stated. The employee should understand that unless unusual extenuating circumstances exist, he will be discharged if he engages in these activities on the job. And he should understand equally well that the rules will be uniformly enforced. A typical listing of prohibited job activities may include the following:

- possession or use of drugs or alcohol
- possession or use of weapons
- being under the influence of alcohol or drugs
- insubordination
- sexual harassment
- theft or possession of stolen goods
- destruction or misuse of property
- making false statements
- excessive tardiness or quitting early
- unexcused absence from work
- refusal to perform assigned duties

- unauthorized departure from work assignment or station
- extreme carelessness
- horseplay or fighting
- disregard of established safety rules
- disregard of established security rules
- offering or accepting graft or gratuities
- soliciting on company-controlled property

Corrective Action

Despite preventive efforts by management, problem employees sometimes find their way into the employee ranks. Then, only corrective or adaptive action (sometimes a combination of both) remain open to the manager. Corrective action may involve reassignment or termination.

The advent of increased government regulation, however, has introduced one other important step in the process: the need for documentation. No longer can a supervisor simply dismiss a nonperforming employee. Prudent managers have found they must protect themselves and their companies against discrimination claims and meritless or capricious litigation. This is accomplished through careful documentation of the reason or reasons for dismissal. A formal reprimand procedure is the vehicle used by many companies to generate the required documentation. In each instance of impropriety, the employee should be reprimanded, either orally in the presence of impartial witnesses or with a written document, the receipt of which is properly acknowledged. Except in major infraction cases in which the evidence is clear and unequivocal, many companies require three written reprimands or warnings before discharge of the offending employee may take place.

Fortunately for the manager, an employee who is a problem for one reason is probably troublesome in several areas as well, and it may be a more effective solution for the manager to take action on the more documentable deviations. A private investigator hired by one project manager determined conclusively that an electrician's helper had been responsible for the theft of some critical electrical substation parts. However, rather commence an arduous criminal action which, although it would have resulted in conviction, would also have consumed a significant amount of time and been somewhat publicized, the manager decided on a different course. Inspection of his file showed that the employee had been previously reprimanded twice for tardiness. Accordingly, the next time he was late, he was given his last reprimand and discharged, a much simpler solution and just as effective: he was off the payroll.

Also, many job application forms include questions relating to prior medical history, such as "Have your ever been treated for the following medical disorders: headaches, dizziness . . ." and "When was your last physical examination or treatment by a medical doctor?" An employee who suppresses or obscures previous medical incidents, such as injuries, is responsible for having made false statements on his application. He is therefore subject to dismissal if making false statements has been included in the list of offenses for which the employee has been told he will be discharged.

A full discussion of this extremely sensitive labor relations area is beyond the scope of this book, and the reader should consult any authoritive text on this subject before setting up documentation procedures.

Adaptive Action

The story about the project manager and the recalcitrant shop steward is an illustration of adaptive action. When the OSHA law first went into effect, it required all employees working above ground to wear and *tie off* (secure to a stable support) safety belts. It immediately became clear that ironworkers working as erectors on structural steel were not going to comply; they felt it was much safer being free to jump out of the way in the event a beam being hoisted slipped than they did being tied to a column with a rope that would restrict their movements. Faced with so much opposition to enforcement, and rightly so, it was not long before most employers made an exception for this situation. *Adaptive action* means working around a problem so that minimum disruption ensues.

In many situations, an adaptive component combined with a corrective component is the best solution. A general superintendent, sent to construct a dam in a remote overseas location, was greeted by both good news and bad news on arrival. The good news was that the work ethic was excellent and many local residents wanted to work on the project; the bad news was that most of them did not even know how to use a hammer properly. His action was a combination of the adaptive and the corrective: he adapted to the situation by committing himself to using what labor was available rather than trying to import it; he corrected the situation by training the people he hired in the use of carpentry tools.

9

Safe Working Environment

Prior to the establishment of the Occupational Safety and Health Administration by the federal government in the early 1970s, the construction industry was shunned by many skilled craftsmen because it was so hazardous. Many of the great construction achievements of the 1920s and 1930s were also the tombs of many craftsmen. Eleven men died constructing the Golden Gate Bridge across San Francisco Bay; many others died constructing the Holland Tunnel under the Hudson River. Construction of the Empire State Building in New York City, then the world's tallest structure, claimed still more lives.

An expression repeated with axiomlike regularity was the macabre cost-death relationship, "For every million dollars spent, a life will be lost." Many old-timers, believing this, approached their jobs every morning with the fatalistic attitude, "When my number's up, I'll get it; until then, no need to worry about the dangers of construction work because nothing can be done about it." In truth, much could have been done if the proper attitudes, on the part of both craftsmen and managers, had prevailed. But they did not.

In 1952, during the construction of a relatively risk-free, single-level government warehouse in Pennsylvania, on a bright summer's day, a craftsman working on a roofing crew casually and literally walked to his death through a skylight opening in the roof of one of the buildings. He was simply going for a cup of coffee and, on the way, turned to say somethng to one of his fellow craftsmen. Unfortunately, the few steps he took while walking with his head back over his shoulder were his last. He fell 18 feet to a compacted earth subgrade.

IMPROPER ATTITUDES

The direct cause of his death was a broken neck. The real cause was the compounding effect of a string of what we recognize today as improper attitudes with regard to accidents.

First in the string was the craftsman's own attitude, characterized by the following:

> I've been in the industry long enough to know how to take care of myself, and I do: I wear a hard hat to protect my head and gloves to protect my hands. That's enough. Anyway, this is no business for the weak-hearted. You've got to take a few chances to make it in this business, but then that's part of the satisfaction. If I were too cautious, what would the other guys think? Besides, when your number's up, it's up. Why worry?

Next was management's attitude:

> "We're sorry about this terrible accident. We'll send flowers to the funeral; we'll do what we can for the family . . . but, really, it was his own fault—he should have been more careful."

And lastly, the public, as represented by government, had the following attitude:

> "A terrible tragedy, and that's why government has mandated worker's compensation insurance—to help all those poor widows. The deceased's family will get a generous death benefit of approximately $10,000. As for the construction industry, well, everyone knows it's dangerous. It's always been that way. Accidents are just something that nothing can be done about."

VOICE OF THE CRAFTSMAN

In the 1960s, with the emergence of the craftsman as a full-fledged member of the middle class, came a stronger voice in what was happening around him and to him in his industry. He was no longer satisfied with putting his life at risk every time he walked inside the gate of a construction project. He was no longer satisfied working with unsafe equipment and tools. He became vocal about cranes with frayed hoisting cables; about electric cords with torn insulation; about ladders held together with tie wire. He didn't like the idea of working up in the air on a scaffold with only loose boards to stand on and no guardrail to rely on—and he said so.

Most of all, he was tired of working for employers who paid little or no attention to his desire to practice his skills in a safe working environment. How could he ever be motivated to do a good job for his employer under conditions like that?

EMPLOYER INDIFFERENCE

From the employer's point of view, even for those managers and supervisors who understood the demotivating effect that neglect of safety in the workplace had, there was no inducement to do anything about it. They were indifferent to safety improvement. In fact, there was pressure *not* to do anything about it because they knew that handrails, ladders, and tool repair all cost money and would put them at a competitive disadvantage in the bidding marketplace. Also, *motivation* was one of those buzzwords professors used in the increasingly popular graduate business schools across the country; it didn't have any place in the construction business, they thought.

FORCE OF LAW—OSHA

In the late 1960s, open-shop and union-shop labor forces found they were, perhaps for the only time in history, on the same side of an issue: safety in the workplace. Largely due to their combined efforts, the Williams-Steiger Act, setting up the Occupational Safety and Health Administration (OSHA), was passed by Congress in 1970.

It took the force of law to bring about change. The law mandated rules and procedures that literally altered the face of the entire building industry. No longer could laborers be placed in deep excavations without shoring protection; no longer could they be forced (even allowed) off the ground without safety belts; no longer could their eyesight be jeopardized by cutting and grinding operations conducted without special eye protection.

The law also provided for enforcement, and, in the spring of 1971, a force of over 1200 compliance officers took to the field to make job-site inspections. The law provided that they had to be given immediate access to any work site they wished to inspect. In some instances, security guards were threatened with personal citations and fines if they even took the time to obtain a clearance from the job manager; that is how serious OSHA was about carrying out its responsibility.

EMPLOYERS' REACTIONS

Most employers, after some initial grumbling about too much regulation by the federal government, accepted OSHA because they knew that without it, the tragic maiming and loss of life would continue forever. More importantly, they realized that if all contractors were subject to the same rules and costs of compliance, they themselves would not lose any competitive edge—everyone would have to include the same allowances for accident prevention in their bids.

Still, many contractors failed to recognize that the new law had benefits beyond its potential for saving lives and preventing disabling injuries.

SAFETY SAVES MONEY

In time, however, experience under the law showed even the skeptics that a genuine effort to prevent accidents would not increase costs. To the contrary, it would significantly reduce the cost of their projects. They finally realized that safety saves money! And the savings are manifested in some dramatic and surprising ways.

WORKMEN'S COMPENSATION

First, reductions in insurance costs, particularly workmen's compensation, are almost immediate for contractors who install effective safety programs, policies, and procedures.

The premiums an employer must pay for workmen's compensation insurance are determined actuarially for each classification of work to be performed on a given project and are charged as a percentage of payroll. Higher-risk classifications are quoted at higher percentages than are classifications with lower risks. On a typical project, cement finishers might be rated at 6 percent while ironworkers connecting or bolting up structural steel high above the ground might be rated at 15 percent. When all crafts on a project are computed, the average for a complete job payroll may be, for instance, 11 percent.

However, these rates are a kind of list price; that is, contractors with good safety records are given discounts, which are called *experience modifiers*. As the name implies, a contractor with low *loss ratios* (what the insurance company pays out versus what it takes in) qualifies for a discount, whereas one with a history of job accidents gets none and may even be required to pay more than the full rate. For a contractor with a continuing good record over a period of time, the discount might be as high as 40 percent, which equates to an experience modifier of 60 percent. This means that instead of paying 11 percent of his payroll, a contractor with this experience modifier would only have to pay 6.6 percent (11 percent times 0.6). On a project with, for instance, 100,000 man-hours of labor, the savings could easily amount to more than $50,000.

DIRECT COST REDUCTIONS

A second way in which a contractor would see lower costs on a project with an effective accident-prevention program would be in direct cost reductions—both in labor as manifested by increased productivity and also in material costs as evidenced by less waste. This is primarily because the craftsman can devote more thought to the productive operations he is to perform and less to concern about his personal survival.

Consider a building project on which, for instance, the superintendent naively believes he can save money by eliminating cleanup labor. Trash receptacles become full to overflowing but are not emptied; trash spills out all over the floor. Stairways become clogged with empty cartons and dropped conduit boxes. Disused rope, wire, and welding cables are strewn helter-skelter throughout the working spaces. Paper cups and drink cans litter all walking areas because the safety discipline of good housekeeping has broken down. The craftsmen feel, correctly, that if the superintendent were serious about running a safe, clean, and productive job, he would at least empty trash before it overflowed.

The cautious craftsman, as he makes his way to his assignment, has to be more concerned about just making it to the point of the work without tripping over toe-stubbers or getting whacked by head-knockers than he does about what he is going to do when he gets there. The result is that his thought process with regard to work planning is interrupted; he simply does not have thinking time available to plan his activities until he reaches the point of the work: just getting there safely is a major mental task. Trying to avoid being hanged by welding cables or twisting his ankle on a scrap of reinforcing rod or catching his eye on a dangling piece of waste wire requires 100 percent attention. He must wait, then, to resume planning his activity. This, in turn, consumes thinking time that could have been applied to execution rather than planning, with the obvious result of higher labor cost to perform the given activity.

Also, on the job without an effective accident-prevention program, when the craftsman finally gets to the point of his assignment, he must clear out trash and waste materials and move cables, ladders, and haphazardly stored equipment just so he can perform the job to which he was assigned. This takes an additional amount of time and, assuming he is being paid journeyman's wages, costs much more than it would have had a proper cleanup program been in place, with unskilled and lower-paid laborers doing the cleaning up.

Instead of making, for instance, 200 inches of weld per day, the welder, after removing all the obstacles in his way, may make only 100 inches, which means that the operation costs twice what it should. It is as if a second welder were standing beside the first, collecting pay but performing no work.

With skilled wages at roughly twice unskilled wages, even if one laborer were assigned exclusively to the welder with no other responsibility but to clean up in front of the welder, half the welder's wages would still be saved. The cost of cleanup labor on a well-managed, safe project normally consumes only 3 to 4 percent of total labor cost. It is, moreover, an inescapable expense; that is, cleanup *must* be performed sooner or later. Trash and scrap must eventually be removed. It is much more expensive to do it all at once, at the end of a project, than it is to do it during the project. And with the additional benefits that a clean project offers, there should be little question that the clean-project route is the less costly one.

FEWER ERRORS

Experience shows that an unsafe job environment is associated with a higher level of wasted materials and *rework* (doing an operation over to correct a mistake). The causes, again, are the same: the craftsman who must be constantly concerned about some unsafe condition or unsafe act will make more mistakes than the craftsman who has a relatively high degree of confidence that he is working in safe conditions. This is not to say that the well-trained craftsman, before thinking about the technical aspects of performing a given task, does not first consider the safety measures required for its safe execution; but rather, after having considered the accident prevention needs, he should be free to devote his attention to performance of the task itself.

For example, consider the pipe fitter who has been given the job of aligning a 90-degree ell to a 45-degree (from vertical) rotation while working from a scaffold 25 feet in the air. He is more likely to get the angle correct the first time if he knows the boards he is standing on are cleated down and cannot slip than he is if he must continually be concerned that any sort of horizontal force may cause the boards to go flying.

The point to remember here is that skilled operations require concentration, and concerns about survival that prevent the craftsman from concentrating on the task at hand will cause him to make errors. Errors must be corrected. This means that additional time will be expended to produce no additional quantity, which of course means loss of productivity.

LESS TURNOVER

Another aspect of cost reduction on a project with a bona fide accident-prevention program is the savings realized by a reduction in turnover. Some craftsmen will quit because of their reluctance to expose themselves to unsafe acts or conditions, and others will be injured and thus unable to perform their normal craft duties. These craftsmen will have to be replaced if schedule pace is to be maintained. As stated earlier, each time a new craftsman is hired, anywhere from a day on a small project to two weeks on a large one is wasted while he becomes oriented to the new set of conditions. In many cases, the only thing the newly hired craftsman is familiar with is his own skill. He must learn a whole new set of personnel and administrative procedures, such as starting time, pay periods, timekeeping, coffee-break policy, call-in sick reporting, job-site rules, and material or tool requisitioning steps.

He must also find out where the supply rooms, storerooms, tool cribs, toilets, first aid stations, and timekeeping offices are and the names of the various areas on the project and how to get to them directly. The type of project or process involved may be new to him, which may mean a change in welding procedures, building-code compliance requirements, or fit-up tolerances; or may be an expansion to an existing operating facility for which special precautions must be taken. For example, in a grass-roots situation, it might be permissible to leave an extension ladder out overnight, but in an operating facility, it might have to be put away at the end of the day to prevent endangering operating personnel in and about the work area at night.

These conditions—a new project, in a new area, for a new employer and building a new type of construction or process—all have a cumulative influence in delaying the new employee from being fully effective for the period of time described above.

And all the time that the craftsman is not fully effective, lower productivity will be experienced: the craftsman is being paid full wages but is not producing the full quantity of work regularly achievable. Cost per unit is thus higher than it would be in the normal environment.

It therefore is prudent for the manager to keep turnover from all causes to a very minimum, and in the case of safety-related turnover, achievement of this goal will provide the multiple benefits described above.

SAFETY TRAINING

An employer who institutes a bona fide accident-prevention program creates, at the same time, many influences that directly motivate the craftsman. For example, even the most basic program includes safety training, both initial and recurrent. The new employee should be provided with safety training as part of his orientation to the project. He may be enrolled in a safety training class that takes place over a period of several weeks and may even be taught first aid, CPR (cardiopulmonary resuscitation), and other skills that will be useful to him at home or play in addition to on the job. He likes this because he is learning something new. Just as in craft training, this new knowledge is an asset provided to him free by the employer and is one he can keep forever. In effect, it is almost as if the manager had given him a cash bonus. He naturally is motivated by the experience and develops a positive attitude toward the employer who has given him this extra reward.

In addition, a manager with an awareness of the benefits of safety training will set up periodic informal meetings among the crews on a project primarily to review and reemphasize safety procedures. They are led by the foreman usually and are held out on the project at the location where the crew works rather than in a classroom. They are sometimes called *toolbox meetings* because, in many cases, the craftsmen group their toolboxes together and actually sit on them during the meeting. The sessions are short, usually no more than ten or twelve minutes, and cover safety problems since the last meeting and prevention of their recurrence as well as special precautions to be taken in the upcoming period. On a normal category project, these may take place once a week; on a particularly hazardous project, they may be held daily.

THE CRAFTSMAN'S REACTION

In any case, they are conducted during the workday, which means that the craftsman is paid while attending. He perceives this—being paid to listen and to talk about factors that affect his health and safety—as a positive sign that management cares about his welfare. Management, in his view, has gone an extra step to create and maintain safety awareness; has gone beyond the minimum requirements of the law to make the workplace safe for him and his fellow craftsmen. He responds by caring more about the job he is doing while, at the same time, working more safely. He realizes that he is needed.

All of this tends to generate a higher sense of self-respect in the craftsman. In addition to feeling needed, which is a basic human emotional need, he feels that he is important enough for someone—in this case, management—to spend time and money through training and attention to safety improvement practices. The act has a positive effect on his self-image.

This is enhanced further when the craftsman sees the manager take steps to provide a clean and orderly workplace. No one likes to work in dirty conditions, especially when they do not need to be dirty. Obviously, digging out footings or laying pipe in a muddy trench constitute working in dirty conditions, but since that is the nature of the operation, the craftsman accepts this when he decides to become a craftsman. What he will not accept are overflowing trash receptacles and scrap materials all over the floor or ground because he knows that all that is required to prevent such conditions is a commitment on the part of the manager to a clean and safe environment. In a more positive sense, the manager who takes the necessary steps is perceived as an enlightened manager.

SAFETY AWARENESS AND SELF-SATISFACTION

Still another motivating influence comes into play with regard to safety awareness in the craftsman. It has to do with the self-satisfaction he derives from achieving goals—in this instance, preventing injury to himself and the people with whom he interacts.

The construction industry remains more hazardous than most industries a skilled worker can work in. Technology and the methods of dealing with dangers have been improved over the years, but it is still a challenge to the craftsman to avoid injury on the job every day, using the accident-prevention skills he has learned and practicing safety awareness. When he completes a day's work without having sustained an injury, he experiences a sense of self-satisfaction. If he has done something to prevent an accident from happening to one of his fellow craftsmen, this feeling of accomplishment is even stronger. The inner satisfaction gained from having achieved a safety goal creates a desire to go back and do the same thing the next day.

It is not a sense of daring that compels the craftsman to go for the goal; rather it is the confidence that having been properly trained in personal accident prevention, he can exercise his skills and knowledge to achieve productive effort safely. It creates the same sense of self-satisfaction that comes to a skier who successfully negotiates the slopes of a steep mountain. The skier's training and skills, when used to convey him safely down the run, develop in him a satisfying feeling when he reaches the bottom. He is motivated to do it again—and do it better.

OBEYING THE RULES

Another type of satisfaction is derived from the completion of the day's work safely. In addition to his own self-satisfaction at having accomplished his personal safety goal, the craftsman derives a sense of satisfaction from complying with the rules set out by the authority figure, who is, in this case, his foreman, superintendent, or manager.

Most people raised in a civilized and democratic environment recognize the need for rules of conduct and strive to obey them. They subscribe to the concept that if everyone obeys the rules, the tasks at hand will be carried out smoothly and safely. Compliance also brings the expectation of a reasonable reward or, at least, avoidance of punishment. In the case of the craftsman, compliance also brings the expectation that he will survive his assignment uninjured and, perhaps, even earn an award for his safe acts or at least be viewed as more promotable. At the very least, he will avoid the premature layoff his less safety-minded coworkers may experience. (Most contractors, mindful of higher insurance premiums, increased threat of litigation, and their moral obligation to keep the workplace as accident-free as possible, will carefully screen out and discharge craftsmen with poor safety attitudes and work habits—for their own good as well as that of the rest of the work force.)

COMMON GOALS

The experience of collaborating with fellow craftsmen to reach organizational safety goals provides a foundation for teamwork in reaching other goals. The collective pursuit of a "safest-crew-of-the-month" award by a carpenter crew, for example, gets them working more closely together, talking about ways and means of preventing accidents so they can win the award. Interpersonal relationships among crew members are thus strengthened and are naturally transferred to other organizational challenges, such as building a particular set of concrete forms in the most economical way possible. The result for the manager is higher productivity, and for the crew it is job satisfaction and the enjoyment that comes with winning teamwork.

EFFECT ON THE MANAGER

A less apparent but nevertheless genuine result of an effective safety program is improved productivity of the manager. The manager of a project with poor accident statistics must, of necessity, spend an inordinate amount of time answering the questions of regulatory authorities, insurance investigators, and attorneys. He must investigate accidents, testify at hearings, and fill out incident reports. All of this reduces the time he has available for performing his main duty, managing the project.

The manager who spends time organizing and implementing a comprehensive accident-prevention program will find that he does not need to waste his time and dissipate his energies on defensive activities such as those described above. Instead, he will be able to take the offensive and pursue the more profitable objective of getting the job done expeditiously. Rather than having to explain to an irate boss why his poor accident statistics have caused an increase in the company's workmen's compensation rates, he may be receiving a bonus for lowering premiums. Rather than having to tell a deceased craftsman's wife that her husband won't be coming home one evening, he will be presenting the craftsman, hopefully in her presence, with a safety award. And if he has any conscience at all, the manager will sleep better at night, knowing he has made the job site a safer place to work.

MANAGER'S COMMITMENT

Instructions for setting up and operating an accident-prevention program are given in several excellent publications distributed by the federal and by state governments and by insurance companies. They should be consulted by the manager in developing his own program. He should, however, keep in mind some basic motivational considerations when doing so.

First, he must convey to everyone involved in his project that he is sincere in his commitment to safety on the job. More than the words he speaks and the notices he writes, the subtle signs and signals he sends out will determine the level of authenticity with which his program is perceived by the craftsmen.

It is vital, therefore, that the manager be seen taking positive action and that these actions be sustained throughout the organization and operation of the program. It is a waste of time for him to talk about implementing a program and then give it only a weak impetus or, worse, start it like a ball of fire and then let it burn out after a few weeks or months. Nothing will turn craftsmen off to management ideas and programs more.

One of the best signals he can send out is to be visibly involved in the operation of the program. He can do this by first selecting an individual as his safety representative. On a project with fewer than 100 employees, the representative might be someone specially trained in accident prevention, but with other responsibilities as well. On a larger project, at least one full-time safety engineer should be assigned. These safety engineers should have no responsibilities other than the administration of the safety program.

After the safety representative has been identified and properly presented to the work force as the manager's direct assistant—his "right arm" for safety—the manager should confirm his support of the program by being seen with the safety representative at least once daily making a safety inspection of the project. These inspections should not be combined with other inspections; they should be performed separately. The manager and the safety engineer should be perceived by the craftsmen as one solid unit. The picture in everyone's mind should be one of the manager and the safety engineer locked arm-in-arm. Thus when the safety engineer counsels an employee about wearing safety glasses, for example, the words should be interpreted by the employee with every bit as much importance as if they had been spoken by the manager himself.

UNQUALIFIED SUPPORT

In addition, the manager must affirm his support of the safety engineer in every instance. His support must be perceived as unwavering. In no instance should the manager allow himself to be placed in the position of adjudicating a dispute between the safety engineer and another employee. The manager must support the safety engineer fully unless he determines that the safety engineer has erred seriously, in which case the safety engineer should be immediately discharged or reassigned. To do otherwise sends confusing signals and jeopardizes the entire effort.

Other signals, perhaps less subtle, should also be sent out. A bona fide first aid station should be set up, publicized, and clearly identified. On a small project, it might be located in the corner of an office trailer and administered by an employee trained in first aid but with other responsibilities as well. Timekeepers or office managers, for example, frequently work well with such a dual responsibility. On a very small job, it may be the foreman or superintendent himself, working with a portable kit stored in a pickup truck or gang box.

MEDICAL STAFFING AND ASSISTANCE

On a large project, a dedicated first aid facility should be set up and staffed with a trained industrial nurse or, in the case of a large project located away from the proximity of a permanent medical facility, a certified physician's assistant. It should be professional in every respect, not only to be fully effective but also to serve as a sign that the safety program is one on which management places the highest priority. On such a project, for example, an exclusive-use emergency vehicle should be maintained in a high state of readiness. (Such vehicles usually retain their value, and the expenditure can be recaptured when the work is completed and the vehicle resold.)

Again, this is only common sense on a large project, but the signal sent out by a first-class vehicle parked in front of the first aid station and ready to go on a moment's notice (with a cadre of trained drivers available) is reassuring to craftsmen for two reasons. First, they recognize that management is genuinely concerned about them, and second, in the event they are involved in an accident, or even become sick or disabled on the job, they will receive immediate help.

Several years ago, almost 2000 craftsmen were employed during peak activity on the construction of a large manufacturing facility in Georgia. Many of the operations in the busy stage of construction revolved around the installation of sensitive production equipment. Fit-up tolerances were extremely tight, and, accordingly, the contractor's most skilled and experienced millwrights were assigned to these tasks. Naturally, many of them were older, and several were past what might be considered normal retirement age.

Unfortunately, one of these older craftsmen collapsed at his work station one afternoon. In almost an instant, however, several actions, proved later to have been life saving, took place.

As soon as the millwright collapsed, another craftsman, who had been given formal first aid instruction in a job-sponsored training class, came to his aid, recognized the symptoms of a heart attack, and began administering CPR. The job-site radio communications network was used to dispatch the project nurse and the emergency vehicle to the craftsman's location. A scheduling engineer located in an office next to the first aid station had previously been given special training in operation of the ambulance. Because of his primary work assignment, he was intimately familiar with the ever-changing layout of the temporary roads and passageways on the project. He knew exactly how to reach the trouble point quickly and drove the vehicle. In the meantime, the telephone receptionist had also been alerted by radio and called the local hospital to advise the emergency medical staff of the inbound patient and the apparent nature of his illness. From a description of the symptoms, the hospital staff confirmed that the craftsman had, in all probability, had a heart attack, and they were ready to treat him the moment he arrived.

The most important result was the saving of a life. The craftsman was resuscitated and, after a few months of recovery, lived many more years and set many more pieces of equipment.

An important additional benefit also resulted, however. The motivational impact on the work force was amazing. The craftsmen realized that a number of people with varied training, responsibilities, and backgrounds had come together as a special team to save the craftsman's life and that the coordinating force behind it was the top management on the project. The incident confirmed to the entire work force that management did, indeed, care about their welfare, and this was one of the factors that contributed to maintenance of high productivity on the project right up to the end of fieldwork. (Many projects experience a sharp drop in productivity as they near the end. This is due to many factors, including loss of momentum as the end objective comes into sight. Unless anticipated and prevented or overcome, this loss sometimes nullifies much of the gain achieved during earlier stages.)

COMMITMENT OF THE WORK FORCE

OSHA requires comprehensive reporting of accident statistics on a monthly or quarterly basis. Included in the report must be the incidence of lost workday cases (formerly called "lost-time-accident" cases), the number of first aid cases, and the number of *doctor's cases* (those requiring the services of a doctor but not serious enough to be a lost workday case). Disregarding this regulation can bring a heavy fine, and so even the most reluctant managers grudgingly comply with this basic reporting requirement. More enlightened managers, however, welcome the generation of this information because it enables them to assess the efficacy of their accident-prevention program and they can use the information to validate their actions.

Even more sophisticated managers go a step further. They use the information to involve the work force actively in the safety program and to obtain its commitment to preventing accidents. This is accomplished by safety training and planning, accompanied by the setting out of organizational and individual objectives for the craftsmen. The statistical results are then publicized regularly via bulletin boards and newsletters, thus creating a level of challenge for individual goal achievement and competition among crafts, crews, areas, or shifts. Incidence of first aid cases, for example, might be focused on as a statistic by which the safest crew of the month is determined, with recognition following as described in Chapter 12. Or incidence of unsafe acts or unsafe conditions, as observed by the safety representative, might be used with equal effectiveness.

The first step is to create in the craftsmen an awareness of the need to work safely, and this is done during the hiring process. It is continually reemphasized at toolbox meetings and in the printed material that is distributed or posted. The second step is to convert this awareness into a positive force that will compel each craftsman to think consciously about the safe way to perform a given task before he performs it. This is more effectively done by *positive reinforcement* (praise and recognition) of the safe acts and achievements than it is by the negative criticism of management.

Criticism by management is received with fear or hostility by many craftsmen. A far better way to get the message across is to make whatever criticism is necessary constructive and to have it come from fellow craftsmen. Once a competitive environment is established, peer pressure will take over and act to bring craftsmen with less than cooperative attitudes into line.

One safety-minded contractor, working under a several-year-long contract for an equally safety-minded major company, used these techniques to establish a national record of over 10 million man-hours without a lost workday accident.

CAVEATS

Only a few caveats with regard to using OSHA-related statistical information are necessary. If the fundamental intent of the job-site program—to protect health and life—is obscured through too much emphasis on simply achieving low incident rates, then the entire concept of accident prevention on the project will eventually fail to achieve its goals. The manager must be sensitive to this possibility and be prepared to reorient both supervisors and craftsmen to the basic purpose of the program.

For example, if a craftsman sustains a serious injury to his back that obviously requires medical attention but is pressured by his peers to avoid seeking that attention, then the program is not working. If a craftsman gets a deep cut but is discouraged from obtaining first aid just to keep his name off the list, then the program is not working, either. In both instances, the medical complications that can result from such neglect are far worse—in terms of the cost, pain, and suffering—than treatment at the onset would have been.

The key is for the manager, both directly and through his safety representative, to watch the safety performance of craftsmen in the field closely and to make sure that their attitudes are appropriate. It is their performance that should form the basis for his planning, and the OSHA statistical information should be used only to confirm the effectiveness of his actions.

10

Basic Personal Comforts

Still another vital element in creating the proper climate for the motivation of craftsmen is the level of attention given to the basic personal comforts. This element, when not properly attended to by management, is significantly more devastating as a demotivator than it is enhancing as a motivator when properly treated. The reason is that craftsmen consider management's attention to their basic personal needs a direct indication of its concern and respect for the craftsman himself. It serves as the master link in the interpersonal relationship chain between management and labor. It is an element that describes the quality of the relationship between the two, and it is therefore continually being evaluated by the craftsmen. The manager should expect little praise from the craftsman for providing him with basic personal comforts at the normally expected level. Conversely, he should expect severe criticism—criticism which may not surface until the damage to the relationship is irreparable—if he fails to live up to the standard.

SIGN OF RESPECT

If the sum of the signs and signals the craftsman sees and perceives are positive, then he feels he will be or is being treated with respect. It is important to emphasize that the craftsman's ego is not particularly fragile—he is not, typically, a prima donna—but he does want to be treated civilly. As discussed in Chapter 6, he's been raised in the era of television, and he sees all races, colors, and socioeconomic levels of people treated equally in situations on the television screen. He wants to believe that what he sees is fact, not fiction, and is customary, not an exception.

LIFE IMITATES ART

Pressure from tens of millions of workers in this country demanding the kind of treatment they have seen on television has caused life to imitate art. The craftsman sees people like himself in situation comedies, commercials, and other scenes to which he can relate, and he begins to consider the clothes they wear, the automobiles they drive, and the way other people treat them as the norm for himself, also. He considers himself an equal and believes that for him to be treated as any less is an act of discrimination. Such an act on the job site is a serious demotivator in itself and has a negative effect on his productivity.

NORMS

Today's office personnel, manufacturing workers, and all others who work in permanent buildings have had the benefit of the basic personal comforts for so long that they expect them the way they expect electric lighting and air conditioning. They have cold, pure drinking water close by; they have clean, sanitary toliet facilities; they work in well-lighted, temperature- and humidity-controlled, noise-attenuated office areas; they have instant communication capability to practically anywhere in the world. These conveniences are the norm for the office environment today in the United States. Only old-timers remember that as recently as thirty years ago, most offices in the nation did not have air conditioning, and the temperature of the water coming out of drinking fountains matched the surrounding air.

Advancement in the construction industry has always lagged behind that in the office and manufacturing environment. In the past twenty-five years, though, construction has progressed at a somewhat faster pace, due to the television phenomenon previously explained and because it started from further behind.

The WPA (Works Progress Administration) crews that built many of our roads and streets had to carry portable outhouses with them. If they were working in an area with sanitary sewers already in place, they removed manhole covers and set the outhouses over the manholes; if they were elsewhere, they simply dug holes in the ground and placed the outhouses over them. Also, drinking water was furnished by a roving *water boy*—the lowest-ranking employee on the project. He spent his entire day filling an ordinary pail up at a fire hydrant or other water source and carrying the pail from crew to crew, with one dipper in it for everyone to drink from. At the same time, even in the most primitive manufacturing plants, inside plumbing and running water were provided to all workers.

There has always been a need for basic conveniences on the construction project, but, in the past, this need was more or less ignored by management. There were several reasons for this: (1) a construction project site is set up for a relatively short period of time; (2) most construction is open to the weather, at least in the beginning; (3) yesterday's craftsmen did not demand comforts to any degree; and (4) the manager spent little time thinking about their needs.

Fortunately, this inequity is disappearing due to the emergence of the new craftsman, who is much more assertive than his father and grandfather were. When it comes to the basic personal comforts, he has reversed the traditional management-craftsman role. In the current relationship, it is the craftsman who makes the decisions about what is acceptable. There are some conveniences he will forego, but there are others he must have.

BASIC PERSONAL COMFORTS

Basic personal comforts is a collective term that describes the personal conveniences the employee has come to expect in today's working environment. Items such as the following should be considered by the manager in evaluating the quality and extent of the personal comforts he provides:

- drinking water
- toilet facilities
- weather-protected eating areas

- site ingress and egress and parking
- telephone access
- food and drink dispensing
- rain gear

Drinking Water

Each craftsman should have immediate access to drinking water at all times. The water should be fresh, cooled in hot weather, and stored in sanitary containers. Disposable cups are part of the requirement, as is a trash receptacle for used cups. Water should be replaced as required but at no longer intervals than once daily. Trash receptacles should be emptied daily because, in addition to cups, empty food containers and food scraps are often discarded in these receptacles. Letting trash accumulate not only presents health hazards but also invites insect and pest infestation.

Many contractors find that a 5-gallon water container will suffice for each crew, with the ratio being varied with the temperature of the working conditions encountered. Ice should be placed in each container when it is filled if temperature conditions warrant. In instances in which the workplace is in an area served by an existing potable water supply and low-voltage power, drinking fountains might be more economical. If the work moves from point to point over the duration of the project, however, the need to relocate the drinking fountains, coupled with the cost of maintaining them, might erase any significant savings.

If a manager gets complaints about the drinking water, he should be concerned. If the procedures for providing this simple comfort are inadequate, it is likely that far more severe problems exist, and he would be well advised to conduct a personal survey of the other basic comforts.

Toilet Facilities

Portable chemical toilets are most commonly used on a project to provide the basic toilet facilities. They are placed at locations around a project close to the work areas so as to minimize the travel time to and from workstations. One unit for each twenty craftsmen is usually adequate, but separate facilities should be provided for men and women. They are relocated as the progress of the work carries the work force to different areas. This type of facility is available on a rental basis through vendors in virtually every section of the United States, and service includes moving the units from point to point as well as cleaning and maintaining them.

If large holding tanks are installed, or if sanitary sewerage exists in the area, more sophisticated facilities can be installed. These consist of small trailers fitted out with multiple stalls, flush toilets, and washbasins. They are also easy to heat and light if electricity is available. This type of facility lends itself very well to a larger project in which a central location will always be convenient to a sizable portion of the work force. Work in existing facilities, such as expansion or shutdown projects, sometimes provides opportunities for the work force to use the permanent toilet facilities. This should be carefully coordinated with management of the existing facility so as to assure that no disruption or conflict between the two work forces will be encountered.

Regardless of the type of toilet facility selected, it should be remembered that keeping the units clean and properly maintained is vital to the success of the craftsman-manager relationship. In the case of chemical toilets, access by the pump truck must be maintained in all weather conditions, so it is usually best to place these toilets near the roads and haul routes on the project. Each unit should be washed down and pumped at intervals that will prevent unsanitary conditions and odors from developing. This may mean once or twice a week in cold weather or every day in hot weather. Also, remembering the need to empathize, the manager should check the condition of the job toilets frequently and not expect craftsmen to use them if he would not do so himself.

Weather-Protected Eating Areas

Many projects, by their type or layout, provide adequate areas for craftsmen to eat their meals in weather-protected space. If the opportunity exists, craftsmen will find a way to take their lunch breaks inside a building or under a canopy, and they prefer it that way—improvising a degree of comfort. Sometimes, however, the design or staging of the project does not permit this, and then adequate eating areas must be created. Crew shacks can provide this as well as a means of storing personal tools, clothes, and safety gear. Crew shacks are usually set up on a craft basis, with at least one for each major craft on the project—ironworkers, pipe fitters, carpenters, and laborers, for example. In cold climates, they can be heated and lighted so that craftsmen can warm up at lunch and change their clothes at the end of the shift, if they wish.

Another option for the manager is to provide a multipurpose area that can be used for large meetings—new-hire orientations, safety demonstrations, and training classes, for example—as well as for eating. These multipurpose areas are normally fitted out with tables and benches, food- and beverage-vending machines, heat, lighting, and drinking water. The area should be kept clean by requiring each person to clean up after himself and by performing a general cleanup and emptying the trash daily.

Site Ingress, Egress, and Parking

Most craftsmen are immensely proud of their vehicles, whether they be vans, sport coupes, four-wheel drives, motorcycles, or pickup trucks. In many cases, their vechicles are, indeed, their most valuable assets. They do not take kindly to having their doors banged up by others because parking spaces are arranged too tightly, and they do not like having their batteries stolen and stereos ripped off while they are at work. They want their vehicles to be safe while they are on the project. It is up to the manager to provide adequate parking at the job site, and the criteria should include the following whenever possible:

- delineated full-size slots with curb bumpers or stop logs
- paving or crushed stone to prevent mud-splattering in wet weather
- dust control
- adequate entry and exit lanes
- security

Slot delineators and curb bumpers facilitate maximum safe use of the available space and provide the driver with a sense of security about the vehicle. A craftsman knows, from experience, that if he parks within the limits of the slot, his vehicle is relatively safe from damage by an adjacent vehicle. In addition, parking-slot delineations usually allow twice as many vehicles to be parked within a given area.

Paving, crushed stone, and dust control are all methods used to minimize the amount of dirt deposited on and in the vehicles. Many contractors provide this service inexpensively by using future permanent roads and parking areas for temporary construction parking. They install the grading, drainage, and subbase materials to permanent specifications to provide employee parking for most of the construction duration. Then, when the project is drawing near to completion and employee work-force levels are being reduced, they relocate the parking areas to already finished areas, regrade and repair, as necessary, the previously used subbase, and complete the final paving.

When this is not possible, the empathizing manager grades off a sufficiently large area—taking care to provide proper surface drainage—and places and compacts one or more courses of crushed stone to stabilize the surface. This provides a relatively durable surface for parking and requires only minor dressing and repair as the project progresses. Mud and ruts in roadways are easily prevented during wet weather using these techniques.

Dust control during dry weather is just as important, if not more so. Stipulations in permits issued for most projects today require that the contractor provide effective dust-control measures. This requirement is usually met by applying water frequently to all unpaved areas subject to heavy traffic during the day. On a well-run project, the manager makes sure that all unpaved temporary access and exit roads from employee parking areas are sprayed by the water truck immediately before starting and quitting time. Two obvious purposes are served by this action: (1) it prevents dust from possibly blowing to off-project locations and creating a public nuisance and (2) it is clear statement to the craftsmen that the manager recognizes the need to prevent dust from getting on and inside their vehicles as they enter or leave the site. Nothing will turn off a craftsman's motivation faster on a hot, sticky day than having to fight his way out after work while being covered and choked with dust from the vehicles in front of him.

Project entrances should be designed so as to permit safe and rapid exit from and entry to the nearest public thoroughfare. The manager should remember that virtually everyone on a project arrives and leaves at the same time. Of the two movements, leaving is far more important to the craftsman—he is on his own time then—and he appreciates being able to get out of the parking lot without delay. Moreover, he is largely unforgiving when management has not expended the effort to make his egress easy.

In most instances, the craftsman is out of sight of his vehicle while on the project. Neither the craftsman nor his employer can afford to have him preoccupied with the security of his vehicle during this time. If the parking area is subject to access by the public, special steps should be taken by the manager. These might include installing closed-circuit television, hiring a roving guard, or both if theft becomes a major problem, as it frequently does in projects situated in densely populated areas. If the parking area is within a project-perimeter fence, the problem is not usually so severe. Most craftsmen respect each other's property, and informal surveillance from a nearby office-trailer or similar point of observation is usually sufficient to deter theft. This is not always the case, however.

A project manager once decided that it would be cost effective to hire a temporary night shift to perform certain operations that required a large work area that was fully occupied during the regular workday. Accordingly, he hired about 100 craftsmen for this second shift. He anticipated that a potential for theft existed because the parking areas were dark. Because the night work was not expected to continue for more than a month or so, the manager decided to increase security by assigning an additional guard to rove the parking lot rather than add lighting. All was quiet for several nights, and then, all in one night, three batteries and two stereos were stolen. Subsequent investigation indicated that, at best, the guard did not cover his post adequately and, at worst, he was part of the theft ring. He was replaced. Even though reasonable precautions had been taken to provide security, it was breached. However, the manager acted promptly. The stolen items were replaced with equal or better equipment, and the guard service was upgraded. The end result was that the confidence of the craftsmen in management was actually increased by the corrective action. No further thefts were experienced.

Telephone Access

Craftsmen have as much need to make personal telephone calls as do managers and office clerks, but the latter have much more conveniently located access to telephones. They use their desk telephones to check up on sick relatives, make dental appointments, and confirm their bank account balances. Craftsmen need to use the telephone for the same purposes. If it is not practical to allow them to use office phones, then coin-operated telephones should be installed at locations that will permit their use as required. If the job has a vending area or canteen, this is the most likely location for such an installation because it allows for their use during breaks and lunch and also provides a place to get change. Alternate locations include outside an office or timekeeping trailer and at the main entrance gate. Care should be taken in selecting a location so as to avoid excessive installation costs as well as vulnerability to vandalism and theft.

Most supervisors recognize that craftsmen need to make personal calls during working hours because the business hours of many establishments coincide with working hours on the project. Since there is generally a rule that a craftsman may not leave his work assignment without the knowledge of his supervisor, telephone use is abused far less in the field than it is in the office.

Food and Drink Dispensing

Vending machines and lunch wagons today compete with the old-style lunch box and thermos in the construction work force. This is due primarily to their convenience. They make hot food and drinks available in cold weather and cold food and drinks in hot weather. They offer a broad range of food, and the food does not have to be prepared at home. In the old days, the man of the family worked. His wife, in addition to keeping the home, made his lunch early in the morning and cleaned his lunch box when he came home in the evening. Today, she also works and leaves the house at the same time (or before) he does; neither has the time to prepare a boxed lunch, and thus a purchased lunch is virtually as necessary as it is convenient.

For the manager, providing some form of food vending on the site is a major advantage. It allows those who purchase their meals to remain on site during the lunch break. Experience shows that this results in less lost time in starting back to work after lunch. Today, because of the food-vending services available, the construction industry has been able to gain the same benefits heretofore available only in more permanent situations, such as offices and factories. And fortunately, the popularity, quality, and variety of vending-machine food and drinks has advanced to the point at which a wide range of wholesome and appetizing meals are available.

A number of different food-vending schemes are available to a typical construction project. Machine vending provides a variety of hot and cold food—sandwiches and soup as well as cold soft drinks. This method remains viable so long as the prices are at least competitive with prices for the same items off the project, the machines are serviced daily, and they offer types of food and drink popular with the work force. Additional benefits can accrue to the work force if the food-service contract is properly structured and controlled. Most vending companies pay a commission for the exclusive right to place their machines on site. Managers with integrity accumulate these commissions and use them to sponsor periodic projectwide events. For example, when the project safety record has reached a significant goal, such as a million man-hours or a full year without a disabling injury, a catered lunch might be set up for all employees to celebrate the achievement. The same might be used to reward successful achievement by the project team of a difficult schedule or cost goal. Skeptics might say that the commission money should stay in the employees' pockets (through lower prices) to start with and that, therefore, management isn't really doing anything special for the work force. The fact is, however, that in addition to the cost of the catered meal itself, many other expenditures come directly out of management's checkbook: the administrative time involved in setting up the affair and the costs of cleanup afterward. More significant, however, is the fact that such lunchtime events usually last at least an hour and a half, and the employer absorbs the payroll time for all but the normal half-hour lunch.

Lunch wagons provide the same type of food and drink service as vending machines. Typically, a food service company will be awarded the exclusive right to bring its dispensing van on site at specific times during the day; e.g., at 9 A.M. for coffee break, 12 noon for lunch, and 3 P.M. for afternoon break. Because they are mobile, these vehicles can make several stops around the site if it is large and bring the service directly to the craftsmen. This tends to minimize the time lost in the break because the craftsman doesn't need to travel far from his workstation. Also, these wagons usually offer a wider choice of food and drink than do vending machines and can be restocked in between trips to the site. The commissions on lunch wagons can also be used to fund project events.

A potential hazard of which the manager must be mindful is that the frequent and repetitive presence on site by outsiders can invite drug trafficking. (This is also true for other outside vendors and suppliers who enter the site perhaps several times a day, such as concrete transit-mix, lumber, and contractor supply company drivers.) An alert manager will have no trouble anticipating the problem and taking preventive action.

Improvisation is sometimes necessary if the manager wants to provide food and drink dispensing. One inventive project manager, building a power plant on a relatively remote site in northern California, was faced with the unavailability of vending service and only one small restaurant within practical lunchtime commuting distance. There was not enough activity in the area to warrant servicing by a regular lunch-wagon vendor, so he devised an alternate plan. Taking full advantage of the only asset that was available (the restaurant's kitchen), he arranged for the restaurant to deliver meals and soft drinks to the site for the duration of the project. Each morning, a restaurant driver brought coffee, rolls, and doughnuts to the site at break time and sold them to the craftsmen from the back of his station wagon. At the same time, he took orders for lunch. After the break, the restaurant staff prepared the meals—most of which were hot due to the cold, damp climate in the area—and the driver returned promptly at noon to distribute them. Prices were the same as in the restaurant. According to the craftsmen, the food was better and the menu offered a wider choice than the lunch wagons they had used in the past on other projects.

Rain Gear

Most outside construction work ceases in rainy weather. Some operations, such as setting steel, are too dangerous when it is wet, while other operations, such as trenching, are slowed down too much by the rain. Other operations, however, once started, must be completed—rain or no rain. Some concreting operations, such as walls that must be poured out to prevent cold joints, fall into this category. In addition, wet weather itself makes several outside operations necessary. Quite often, temporary drains and water diversions must be constructed as the rain falls, and recently completed work must be protected. Occasionally, severe weather damages previously installed protection or the work itself, and emergency remedial measures must be taken.

Craftsmen customarily provide their own warm clothing, but they should not be expected to provide special protection gear. Those assigned to work in inclement weather, for example, should be provided with adequate rain gear—hats, coats, coveralls, and boots—in addition to any special safety gear they need. Also, those working in wet holes and with their feet in concrete should be provided with rubber boots, for obvious reasons.

REASONABLE EXPECTATIONS

None of the personal comforts described above are unreasonable when viewed in the context of today's society. The craftsman wants, after realistic allowances for the condition in which construction is carried out, what everyone else in the working environment wants. He wants what experience has taught him can be provided by management through readily available and inexpensive sources. For instance, although he might wish for an assignment in a warm, well-lighted building all the time, he expects to be out in the elements most of the time. This is acceptable to him because he knew the construction industry involved outside work and, after all, it was his decision to make a career in it. When he encounters only what he expected, he's satisfied.

The degree of personal comfort that will satisfy him varies according to the type of construction project. The craftsman doing the initial layout on a remote site obviously expects far less than the craftsman who will work on a site after the project has been mobilized and the temporary facilities have been put in place. The craftsman on the remote site may be 10 miles from the nearest gas station and have to drive that far each time he needs to make a phone call or fill his water jug. He may have nowhere to go except the cab of his truck in the event of a surprise windstorm or rain shower.

On the other hand, the craftsman hired on a project already under way expects, for instance, portable toilets, drinking water, and at least a pay phone nearby so he can call home if he's going to have to work late. Further up the degree-of-permanence scale, the craftsman working as a welder in a fabrication shop, where there is obviously electric power and running water, expects indoor plumbing on a par with the plumbing in any other manufacturing operation.

The key point here is that the craftsman knows approximately what to expect before he decides to accept a particular assignment, and as long as the level of comforts conforms to this expectation, he's satisfied.

GOING BEYOND THE BASICS

The craftsman also expects, however, that management will do its best, given site conditions, to provide the conveniences. He is particularly pleased when he suggests improvements beyond the basics that do not add cost and management acts on his suggestion.

On a project in the mountains of Colorado several years ago, it was necessary to construct foundations in the winter to keep the job on schedule. That meant that large timber and plastic enclosures had to be constructed over the excavations, and the interiors had to be heated to keep the ground and concrete from freezing. The daytime high temperature was reaching only 20 degrees above zero, and the wind chill was down around minus 20. Management had placed portable toilets around the project in accordance with its standard procedure, but the craftsmen came up with a suggestion that improved the situation considerably. The idea was adopted, and, with very little extra effort, they relocated the portable toilets into the walls of the heated enclosures, with the doors on the outside so everyone could get to them but the main part of the toilets on the inside, where the temperature, while still not what would qualify as comfortable, was nevertheless much warmer than it was on the outside. The end result was enhanced personal comfort for the craftsmen with an extra motivational benefit due to the fact that it was their suggestion, positively acted upon by management, that made it possible.

DEGREES OF "ROUGHING IT"

When he chooses the type of construction he will work on, the craftsman is also making a choice with regard to personal comfort level. A carpenter trainee coming out of vocational training has choices to make: he can opt to go into a shop and make cabinets, or he can choose to go into field construction and build houses, and timber structures and concrete forms. When he makes his choice, he is determining both his level of pay and the conditions, including the level of basic personal comforts, under which he will work.

Of course, when he chose to go into construction, he was already accepting a lower level of personal comfort than he would enjoy were he to have chosen manufacturing. In his mind, that's acceptable because he has decided, all things considered, he'd rather be in construction.

Also, the craftsman recognizes that because of the relatively short duration of a construction project, neither the quality nor the quantity of the basic personal comforts can reasonably be expected to be as good as those in a manufacturing environment.

On the construction project itself, there are sometimes differences in comforts between the field and the office. The craftsman is aware of these differences and, unless they are extravagant, accepts them. This is because he acknowledges the advantages of heat and air conditioning for people who sit at desks all day doing paperwork. He obviously also acknowledges the need for running water and inside plumbing for this group of people. He may even take a somewhat macho pride in comparing himself to the office staff, picturing himself enduring adverse conditions when they do not.

To some, in addition, there is a certain glamour in the construction industry. Accepting lower levels of comfort may, to some degree, enhance the self-image of the craftsman and contribute to his ego satisfaction. "It was really rough out there, but I handled it." Indeed there is some substance to this concept.

For example, a welder in manufacturing today enjoys, with few exceptions, an environment in which the temperature, humidity, lighting level, and pollution level are closely controlled. In contrast, a welder with the same skills in the construction industry may work in south Texas in the summer in 95-degree heat and 95-percent humidity or in Wyoming during the winter in a rarefied atmosphere 6000 feet above sea level with a 20-degree temperature and a constant 15-knot wind. However he accepts this because it's part of being in the construction industry. He knew he would be working outside when he made the choice to learn a field-construction skill.

CONTROVERSIAL COMFORTS

Some comforts are controversial, and providing them should be analyzed carefully for all employees on the project. The need to be even and consistent in setting job policy, as described in Chapter 6, should be taken into account. Typical of the questions to be answered is whether coffee, snack, and cold drink machines should be provided for office employees when they are not available to craftsmen in the field. Again, empathy by the manager will help answer the question.

It is fair to permit office clerks, many of whom don't have skills equivalent to the skills of craftsmen in the field, to sip coffee, tea, or soda and eat crackers at their desks at any time of the day and deny craftsmen the same privilege? The rote response to this question in the past has been that such activity doesn't interfere with the clerks' productivity because they can continue working while eating or drinking; if craftsmen did the same thing, it would interfere with their assigned tasks.

A closer look at this situation is revealing, however. Can someone write or type while holding a cup of coffee any more easily than he can drive a nail with a hammer? Can a person communicate on the phone with a mouth full of crackers any more easily than a welder can lay an acceptable bead with a doughnut in his hand? It is not really a question of capability but of efficiency.

Many times craftsmen are not permitted to do these things because the manager who sits at a desk and drinks coffee while working has made a judgment that although such diversion does not slow him down, it would slow craftsmen down. While the craftsman recognizes that he can't do his job efficiently while holding a coffee cup in one hand, he probably feels that the office clerk can't either. The weight of statistical evidence, developed in recent office-practice studies, confirms the craftsman's viewpoint. A recent study estimated that the average office worker actually works at his or her basic job assignment only three hours and thirty-six minutes out of eight hours in a typical workday.

Management thus has no rational basis for denying comforts such as these to the craftsman while allowing those in the office to have them. The decision to do so is therefore subjective or emotional and has to do more with the obsolescent attitude of privilege and rank, also discussed in Chapter 6, than anything else. And this becomes another serious demotivator to the craftsman. People working in the office are no more important than is the craftsman. They are no more vital to the success of the project; in fact, as discussed earlier, they are at their desk to support the craftsman so that he can get his job done. To treat them better with regard to basic personal comforts such as these rarely has merit.

Considering only nourishment needs in basic terms, who is more in need of liquid and sugar on a hot day: the clerk in an air-conditioned office trailer or the craftsman out in the hot sun? Who needs the calories in a doughnut more: the clerk working at a computer with his fingers or the craftsman moving concrete with a shovel?

MOVING WITH THE WORK

The manager must empathize not only when developing the project and planning the basic personal comforts but also when looking at them throughout the project. This is necessary for a number of reasons. First, the project itself is a dynamic operation; things are constantly shifting. For instance, as the work activity shifts from site grading and drainage work to building construction, the center of work activity shifts to a different location on the site. The manager must make sure that as the work areas shift, the support facilities, such as toolrooms, portable toilets, and drinking water, move with the work to new areas, also. This may seem obvious, but it is frequently overlooked by managers during the project planning stages. Sometimes the cost of moving and maintaining these necessities is inadvertently omitted from the job estimate. In this case, the manager may opt to avoid the cost rather than admit an oversight and change the cost projection.

MAINTAINING COMFORT LEVELS

The manager must make sure that the personal comforts incorporated in the original plan are kept current with the size of the work force and that the concepts are maintained as originally designed. It is demotivating to the craftsman to be given certain comforts at the onset of a job and then have them taken away later. For example, if a decision is made to water the dirt roads and parking areas to control dust and this is done at the beginning of a job, the manager must be willing to maintain this feature for the project's duration. It simply won't do for him to cut the watering operations down at some point later in the project because his overhead budget is under scrutiny by management or he needs to cover an overrun in another account. The people hired for the project were aware that the roads were being watered when they started work, and they expect to see the watering continued, as necessary, while they're there. If, on the other hand, dust control wasn't in the original plan, those hiring on would have been aware of this, also. And if they made the decision to hire on anyway, they would have expected not to have it, but it would have been their decision. Those who didn't want to accept employment on a project with no dust control would have had a choice to work somewhere else.

TAKING AWAY

Once again, changing the rules halfway through a project, taking away something that previously had been given, is demotivating to craftsmen. They still have a choice, but the choice is now whether to change their standards of acceptance and stay on the job or leave for employment on another project. The good craftsman will maintain his standards of acceptance every time if he can find work elsewhere. If he can't, he'll remain, but in a less motivated condition.

One of the worst transgressions a manager can make is, consciously or unconsciously, to punish craftsmen for some perceived misdeed by taking away their basic personal comforts. After all, the whole idea of providing basic personal comforts is to acknowledge that the craftsman is entitled to basic humane treatment. To acknowledge this at first and then to deny it by removing comforts says to the craftsman that management not only does not care about him but actually views him as unworthy of basic humane treatment and, more importantly, respect. It is perhaps one of the most damaging demotivators a manager can create and will cause a devastating adverse reaction in the craftsman every time. The end result is a negative atmosphere that most likely will not be eradicated before the project is over. In many cases, also, the reputation of the manager is tarnished to the extent that he can no longer motivate a project work force and complete programs successfully. He thus becomes of little further use to his employer.

On a very large industrial plant project several years ago, a project manager, through one simple act, not only demotivated the entire work force but, before it ended, induced almost mutinous reactions among them. It started when he noticed that craftsmen seemed to be going to the field-located cold drink machines almost every day from 10 A.M. on.

Without investigating further, he concluded that too many people were taking soft drink breaks in addition to coffee breaks and decided to take corrective action. He directed that a chain-link fence with a padlocked gate be put around the drink machines and that the gate be opened only from 12 noon to 12:30 P.M.—the lunch period for the project.

The result of this action was to cause long lines for the drinks, and those not fortunate enough to be in the front of the line didn't get their drinks until almost 12:30 P.M. Those at the back of the line didn't get a drink at all. Did those not able to get their drinks until 12:30 P.M. skip it and go back to work? Of course not. They still took the time to eat their sandwiches. This cost the project not only the time lost while the craftsmen ate after the beginning of the afternoon shift but the lost productivity due to lowered motivation. It also caused a direct loss because of increased *rework*—having to perform operations over two and three times because of poor workmanship or errors. The workmen simply did not care about performing their jobs to their previously high-quality standards.

The end result was a loss of productivity far greater than that caused by the previous practice. What the project manager didn't realize or couldn't see was that people were getting their lunch drinks throughout the morning at whatever time they happened to pass or be near the machines and there were no lines of people waiting. They were putting the drink cans in their lunch boxes, saving them, so they wouldn't lose any productive time standing in line or eating a late lunch. In reality, they had been trying to be as productive as possible, and the manager completely turned them off with his ill-conceived action.

SIGNS OF MANAGEMENT QUALITY

Choosing a construction project on which to work is like going to a movie without reading a review beforehand. The craftsman must wait until he is on the job and can observe conditions before he can make a decision on whether or not to remain. He continually looks for signs of the quality of management, and the most obvious signs to him are the quality of the basic personal comforts. He knows what to expect because of his previous experiences in the industry. He initially rates a job and its management when he goes through the gate for the first time. If he likes what he sees, he stays; if he doesn't, he leaves as soon as he can. Each project has its own distinctive personality, which is a reflection of a collective personality of the job's management.

The quality of the basic personal comforts is more of a signal to the craftsman than is the quantity. The craftsman knows that there are time-tested rules of thumb for how many units of a particular necessity are required per employee (quantity), but the signals he looks for are the shape the comforts are in (quality). For instance, he will not be impressed positively or negatively by the number of field toilets, but he will rate the management on their cleanliness and convenience.

Other signs he will immediately draw conclusions from are, for example, the way drinking water is provided. Is each water barrel filled with cool, fresh water and distributed at the beginning of each shift, or is it left in place until it has been empty for a day or so and then refilled? Is there an adequate supply of cups? Is there a trash barrel for disposal of used cups? Is the trash barrel emptied when it is full, or are cups blowing all over the site?

Also, is a weather-protected eating area provided? Is it away from dust and fumes? If it isn't, he will have to eat in his truck. If he is on a time-clock project, that means clocking out and in again, which, in turn, means that the timekeeper must arrange to have the clock station staffed for the lunch period. Even if there is no time clock, there will still be time lost getting back to work, in addition to the potential hazards due to beer drinking or drug use in the parking areas. For these reasons and others, it is much better to provide a lunch area and encourage the work force to stay on the project site during the lunch period.

As a corollary to the lunch question, are adequate cleanup areas provided so that the craftsmen can wash dirt, grease, and chemicals off their hands before handling their food?

What about the exit road? Is it so dusty that they have to leave with their windows rolled up to keep from choking on days when the outside temperature is 90 degrees? Is the layout of the exit road such that they can exit in five minutes or so, or do they have to wait in line for twenty minutes just to get off the project?

What about the time-clock alley, if there is one? Is it secure enough to prevent cheaters from turning in fraudulent time cards? The average craftsman is basically honest, and he does not approve of cheating on time cards even though it is not his money that is being stolen. But he also feels that it is management's responsibility to set up security controls.

And if the time-clock alley is secure, is it big enough to prevent unnecessary delays when craftsmen come in in the morning and, more important, when they're in a hurry to get home after a hard day.

All these things are indications of management's collective attitude and the quality of the operation. None of these items are any more expensive to set up correctly than they are to set up incorrectly. In fact, they're all minor expenses. The expense occurs when they aren't done right initially and have to be done over. A larger cost, moreover, is the loss in real dollars due to the demotivating impact of management's failure to acknowledge the craftsman's need for them.

When it comes to basic personal comforts, "little things mean a lot." Obviously, no craftsman is going to die of thirst on a construction project today; he will find a drink somewhere, even if he has to walk the whole job to find it. What is important is the signal that a lack of convenient drinking water can send out to the craftsman, what it reveals to him about the quality of management. The craftsman asks himself, "If management can't provide drinking water to the quality and quantity levels sensibly required, what's the probability that it can provide a safe working environment, for example? What's the probability it can provide adequate supplies of tools and equipment? What's the probability it can properly coordinate the work so that we can do our jobs?"

CONCLUSIONS DRAWN

When the good craftsman sees too many of the wrong signs, he goes somewhere else to work: the people left on the project are substandard workers who are unable to find jobs elsewhere. Even in times of high unemployment, good craftsmen—those who are vital to the success of a project because they pace the work and set the standards of achievement in safety, productivity, and pride in workmanship, those whom the manager must have on his project if he wants to achieve his cost and time goals—can always find work.

It is human nature to draw conclusions based on the limited visual clues described above. It may not always be reasonable or logical, but most of the time, notwithstanding the means, the end result—the conclusion the craftsman reaches—is correct.

We all do the same thing at one time or another in our everyday lives. For example, how many times have we heard a car buyer vow never to buy a particular make or model of car again because of its poor quality. When pressed for details, the buyer may admit that what really bothers him is a loose window handle, an ashtray that rattles, or some other small item that is not up to his standards. Though it may seem unreasonable, the fact is that that particular automobile manufacturer has lost a customer.

Moreover, the car buyer's decision was probably correct. Eliminating defects such as loose window cranks and rattling ashtrays is easy for the manufacturer to do, and when the pressure from lost sales gets intense enough, the defect will be eliminated.

The reasoning of the car buyer is "If they can't get this right, what about the engine, and the transmission, and the brakes, and the front suspension? Will they last, will they cause me trouble, will they make it an unpleasant experience for me to keep this car?"

It is the same in the construction industry, except it is the craftsman looking for signs. Instead of loose door handles and rattling ashtrays, the signs are a multitude of easy-to-spot things that indicate the attention given by management to basic personal comforts.

For management to fail to provide the proper level of these basic personal comforts discloses a number of deficiencies, including poor planning ability. More revealing is that it shows a low level of understanding by management of the basic procedures necessary for a given project to run smoothly and achieve success. Most revealing to the craftsman, however, is the lack of concern by management for him and his fellow workers. The craftsman will respond in kind in practically every instance by showing little concern for management's objectives, whether they be cost performance, schedule achievement, quality, or safety.

11

Training

Training a worker to do his job obviously will enable him to accomplish the tasks assigned to him more productively. Knowing what to do allows him to work more quickly and make fewer errors than he would without the training. Output is also enhanced because of the motivational benefit of the training experience itself. Being taught the proper way to perform the task permits each repetition of it to become a successful and, therefore, satisfying experience.

Training provided by the employer may take many forms. Each form benefits both the giver and the receiver—management and the craftsman. Management gets a reserve of people with craft and supervisory skills who exhibit an uncommon loyalty to the provider of the training—the employer—as well as to each other, and the craftsman acquires an intellectual asset that leads to job enrichment and improved socioeconomic status.

TRAINING ARRANGEMENTS, COSTS, AND FUNDING

Employers operating training programs usually furnish the facilities, including classrooms, work areas, benches, tools, and equipment. They pay for utilities, maintenance, and janitorial services. They furnish the teaching materials, such as texts, workbooks, visual aids, and equipment. They provide the materials consumed during demonstration and hands-on practice.

Instructors are also paid by the employer, and since, in many cases, the instructors are also full-time employees, the time they spend in the classroom is usually paid for at an overtime rate.

All this is normally at no cost to the trainee, and thus the employer has a sizable investment in each person accepted for training. This makes the selection process vital to the success of the program. The employer must seek out not only those with work habits conducive to absorption of the training but also those whose work ethic and sense of fairness will persuade them to stay with him on the project or projects long enough for him to gain the benefits of their training.

In some cases, the employer can recover part of his cost—usually limited to instructor costs—from the government. Under certain conditions, a state may make funds available for craft training that it determines is in the interests of a particular region. These funds are usually channeled through state vocational training or other educational institutions. Most times, funding is limited to instances in which the work site or reserve of available labor is far removed from the institution, the number of trained workers required exceeds the capabilities of the institution, or the skills required are not offered by the institution's program. Programs vary from state to state, but, in general, eligibility criteria may also include targeting recruitment toward special groups, such as the unemployed or the economically disadvantaged.

TYPES OF TRAINING

A full training program gives the craftsman an opportunity to learn and advance to every job level from craft helper to superintendent and beyond. It provides basic entry-level skills training, upgraded skills training, skills cross training, and supervisory training, as well as specialized training in areas such as safety and first aid.

Basic Entry-Level Skills Training

If an employee is new to the industry, basic entry-level skills training will prepare him to make a positive contribution to a construction effort by teaching him the basic operations in his craft. He can exercise and sharpen them up as a helper while he assists fully skilled craftsmen and learns additional skills through on-the-job training. Perhaps more important is a feature of a well-run program that emphasizes accident prevention and prepares the craftsman to operate in the construction environment without hurting himself or others.

Helper training is usually provided to candidates selected from the unskilled-laborer ranks and from job applicants who, by their past employment history, show initiative and the potential to learn craft skills. It is provided on an unpaid basis; the student does not earn wages for the time spent in training sessions.

Instructors are usually chosen from the cadre of craft supervisors. In addition to ensuring that the teaching methods are directly suited to the sponsoring contractor's needs, this also permits maximum interaction with students and allows the communication process to begin early. Classes are held either during the day or at night depending on the makeup of the student group. If most students are unemployed, accelerated training takes place eight hours a day, five days a week, to provide them with the skills necessary for employment as quickly as possible. If they are already employed, on the project or elsewhere, class meetings might be held for two hours a night, two or three times a week over a longer period. In most crafts in which helpers are employed, the curriculum includes safety, tools of the craft, terminology and nomenclature, materials of construction, and hands-on practice.

Training periods vary from craft to craft, depending on the scope and complexity of skills to be learned. Insulator helpers may graduate after two weeks, or eighty hours, of training, while flat-plate welders may require eight to twelve weeks of full-time instruction and practice.

Ungraded Skills Training

Students selected for upgraded skills training are normally drawn from those helpers who show mastery of the basic skill elements of their craft and are recommended by their supervisors as ready to progress into journeyman training. This program carries the seasoned apprentice through the full breadth of operations in which a qualified journeyman is required to be proficient and prepares him to approach virtually any activity or operation in his craft with confidence.

It involves classroom instruction, supervised study using self-paced instructional materials, some homework, and on-the-job practical application. Class time is usually three to four hours per week. Classes are conducted during nonworking hours, and students are not paid for their time. As with entry-level training, the instructors are experienced craft foremen, general foremen, and superintendents who have been exposed to the basic elements of educational training as well as to supervisory training.

Training periods vary from craft to craft and from student to student. Because much of the training is self-paced, students who start with more experience or greater aptitude for the craft finish earlier. The average student in a civil craft might require eighteen months of this regimen to complete a given course. Students in mechanical and electrical classes may require up to twenty-four months to complete their study.

Certificates confirming that the craftsman has met the requirements of journeyman designation are usually given upon completion of a written examination and a supervisor's recommendation. In the case of welders, successful completion of testing for a specific welding procedure certifies him to weld to that procedure only. Procedures vary by welding type and material, and it is not unusual for an experienced welder to possess half a dozen or more certification cards.

Good managers take advantage of the opportunity afforded when a class graduates and take the time to give genuine recognition for the accomplishment with meaningful ceremonies for them.

Skills Cross Training

Cross training provides a craftsman with additional skills in another craft. Although in some cases it may enable him to advance to a higher-paying craft, the purpose of cross training is to qualify a journeyman in one craft to be a journeyman in another. It is thus sometimes called *lateral training.* A carpenter might take lateral training in the millwright craft, for example, so that he can qualify for a broader range of employment opportunities.

This type of training is important as a motivator because it not only broadens the craftsman's skills and thus enhances his self-esteem but also allows him the option, if he wishes, to progress to a traditionally higher-paying craft and thereby stay at a project location longer.

For instance, consider the craftsman who enters the industry as an ironworker helper. Through *upgraded skills training,* he becomes qualified as an ironworker journeyman. Through *additional upgraded skills training,* he becomes certified as a structural welder. He is then qualified to weld on flat surfaces. At this point, if he wishes to progress further, he can take training in pipe welding, making a total of two upgrade and one cross-training sequences he has taken. After that, he can take supervisory training.

Supervisory Training

Supervisory training is normally first accessed by journeyman-level craftsmen who show qualities of leadership and initiative. These students come from two categories: those who have been promoted in the field due to expansion or attrition of supervisory personnel and must be given the training to strengthen their management skills, and those who demonstrate the potential and are offered the training on a growth and look-ahead basis.

This type of training, instead of focusing on the improvement of manual skills, trains the craftsman student to deal with people. It teaches him interpersonal relationship skills. Like other types of training, it starts off on a voluntary (unpaid) basis and is usually given two nights a week for two or three months. The curriculum includes communications, human relations, psychology and motivation, laws and regulations, planning, scheduling and estimating, cost control, and administrative procedures. It may also include information and background on the company, its philosophies, policies, business outlook, long-range plans, and related material. The relationship between quality, productivity, and competitiveness in the marketplace may also be stressed.

Supervisory training has become acknowledged as a needed element over the past several years in recognition of the fact that even top-notch craftsmen, though they may be consummately skilled in all the operations of their craft, are usually not successful as supervisors unless they have learned the human side of the craft.

There are several levels of supervisory training. The first is an introduction to management and is designed to transform the experienced journeyman into a foreman. Its thrust is to make the individual aware that his is the first line of management and to provide him with the resources needed to act as such. Each subsequent course progresses further into the elements of management so that at the superintendent level, the graduating student will be prepared to handle virtually any supervisory problem and deal with even the most difficult types of personalities.

Higher levels of supervisory training, offered in whole or in part by outside organizations and institutions, may include seminars on various topics. Costs and salaries of this additional training are usually absorbed by the employer because by that point, both the supervisor and the employer are comfortable in their relationship and willing to invest in each other.

Training obviously prepares students to produce units of consistently acceptable work and thus to work more productively than untrained craftsmen. Beyond this, however, training has an intense motivational impact.

THE SATISFACTION VALUE OF TRAINING

Training is a learning experience. As such, it possesses all the characteristics of other human learning experiences. A baby who has learned that moving all four limbs in coordinated motion results in a new mobility called *crawling* is positively exhilarated by the experience and shows his exhilaration with great glee, as any parent can attest. Not only is he overjoyed at having accomplished something new, he is also more confident when it comes to trying something even more challenging—walking.

Similarly, everyone remembers their first successful ride on a bicycle without training wheels or a parent's steadying hand. The self-satisfaction of that experience is exceeded only by subsequent learning, such as driving an automobile. The engineering student is pleased with himself when he realizes he has gained the capability to design a bridge upon completion of a course in structural design. He derives a large sense of satisfaction just knowing that he possesses the skill to perform that task.

Learning is a positive and ego-reinforcing experience, whether it is learning a physical skill, increasing one's knowledge, or a combination of both. So it is with the craftsman. In fact, the craftsman is doubly pleased with himself upon finishing a job-training course because not only has he learned a new skill but the skill he has mastered will increase his earning power.

For many, the training experience brings still a further satisfaction, one rooted in the concept of challenge and reward. Typically, because of the potential opportunities for himself he sees growing out of a particular training program, the craftsman has a strong desire to participate. He decides to undertake the training because he believes he can complete it successfully. Also, he considers it a challenge. Nothing makes him more pleased with himself than learning something new, trying it, and finding that he can do it. It at once exhilarates him and motivates him. In addition, the confidence that comes from learning how to perform a task and then performing it makes him feel that he can perform even more advanced and complex tasks.

Psychologists call this the *self-fulfilling prophecy.* If the craftsman believes that he can perform a task, the chances are that he will be able to perform it. Conversely, if he believes he cannot, indeed he won't be able to perform it.

ADDITIONAL BENEFITS

A young man with a few years' experience as a house-framing carpenter but no experience in heavy construction once went to work on a large industrial plant project as a carpenter's helper. His house-framing experience qualified him to work in a concrete forming crew because both tasks involved nailing dimension lumber. One day, while in the timekeeping office, he met a secretary and was smitten at once by her beauty and charm. A typical boy-girl relationship ensued—for a while. The helper realized one day that as soon as the concrete work on the project was finished, he would be laid off, and that event would most certainly jeopardize his romance with the secretary.

Then he read on the bulletin board that a training class in pipe and equipment insulation was about to begin. He quickly applied for the class—it was going to be held in the evenings after work three nights a week—because he realized that the project was moving into its mechanical phase and he would be able to stay on the job several more months if he could get transferred to an insulation crew. His strategy worked. He finished the course with praise from his instructor and was put to work as an insulator when the carpentry work was finished. The romance continued to flourish for several more months until he realized that the insulation work would also finish up in the near future. Then he learned that a training class in industrial painting was about to begin. Again he went to school three nights a week and completed the painting course with flying colors. When the insulation work was finished, he moved over and began painting.

Predictably, the end result was "boy marries girl." Just as exciting from a career viewpoint, however, the young man had learned three new skills. Perhaps more important, he had demonstrated to management that he was not afraid of hard work and had the initiative (whatever his motive) to learn new skills. He and his new bride moved on to several subsequent projects with the contractor, and, before long, he was selected to begin supervisory training.

He remained with the same contractor for several years and progressed upward through the supervisory organization. He eventually had his own projects and was one of the contractor's best superintendents. Training provided him with the means to progress. It also taught him many subtle lessons in the value of the learning experience, challenge and reward and the self-fulfilling prophesy. In his case, each of these concepts had real meaning because he knew that if he hadn't approached the challenge of learning a new skill positively and with the motivational drive to put forth the extra effort required, he would not have succeeded. And if he had lost out in the training, he would also have lost out in love.

Much of his later training, of course, took place after his marriage, and his dedication obviously sprang from motives other than simply his pursuit of love. It is not clear just when this craftsman's priorities switched to self-improvement, but his initial desire for knowledge, skills, and advancement was at least kindled by the first courses he took.

SHOWING CONCERN THROUGH TRAINING

Perhaps the case of the lovesick craftsman described above is not typical of the drives that make craftsmen want to take the time for training. Most of them more likely are motivated by a desire to improve their socioeconomic status and attain higher esteem.

But there is also a motivating element that involves the relationship between craftsmen and supervisors. By putting forth the effort to taking training, the craftsman sends signals to management about himself, all of them positive. Also, by spending the money to offer training, management sends signals to the craftsman about itself, and again, the signals are positive.

Employers rarely commit themselves to spending a lot of money on training programs when skilled craftsmen in each craft are projected to be readily available for some time. That's just simple economics, and everyone understands it. However, once a need becomes apparent, the company that looks ahead and commits funds to set up and activate a training program is sending out important positive signals to craftsmen. It effectively states that it is proceeding on its own to prepare itself and its employees for future growth. It is saying that its future looks bright enough to justify spending the time and effort required to get such a program up and running so that when growth level demands it, it will have a staff of fully trained craftsmen and supervisors to carry out the work.

In general, people like to associate with a successful enterprise, and craftsmen are no exception. When they see that a company has a training program in place, they associate this with success and are drawn to the company regardless of whether they themselves need or want training.

Perhaps most important of the signals a company training program sends out are those that are interpreted as being beamed directly at craftsmen in a personal sense. It is these signals that create in the craftsman's mind a picture of the company's corporate personality. In the case of training, he concludes that beyond the benefits the company itself derives from training, it also cares about craftsmen and their welfare.

By providing training, management is, in effect, saying, "We are concerned. We care about you. We want to see you improve your standard of living. We want to see you progress further and higher in the construction industry." This is an intensely personal message, and craftsmen perceive it as such. Of course they recognize that management benefits because the craftsman

similar to what one might encounter in a small school or college. In the theory part of the class, an informal question-and-answer environment is created. In the hands-on-tools (or laboratory) part of the class, the instructor moves from student to student, demonstrating the how-to aspects of the instruction. Both situations provide an excellent atmosphere for development of close relationships between student and instructor. Close relationships in turn create communication levels that may bring out problems, attitudes, gripes, and suggestions for improvement that would otherwise remain undiscovered, suppressed, or hidden. The information obtained by the instructor is in addition to what the manager may learn on his visits because of the continuing close contact between the craftsman and the instructor for two hours a night, three nights a week, or whatever the setup is.

The craftsman-instructor relationship is somewhat more intense than the relationship between the craftsman and the manager. The instructor most likely has a craft background himself and is with the craftsman more of the time. The tie is viewed by the craftsman as more like an older-brother relationship, while that with the manager is more like a father relationship. Naturally, the craftsman will confide more in the older brother and will, perhaps, tell him more of the negative things he sees on the job than he would the manager. All of this is healthy so long as the right amount of feedback is communicated to the manager so he can correct the problems uncovered.

TRAINING IMPROVES METHODS

An additional direct benefit of training is the fact that the manager has on his project a group of craftsmen with improved skills who will install the work. This means that the work will be carried out using methods and techniques that the employer considers best. There is also a higher potential for eliminating rework because of the improved skill levels being employed. Still a further benefit is for the manager to monitor closely the post-training craft operations and improve them even more by fine tuning the methods and techniques that have been taught in training classes.

Using this method and capturing the action with a time-lapse camera, one employer was able to reduce his concrete placement costs by 35 percent. In this instance, he was able to gain the full support of the concrete crew because he let them review the tape. In addition to studying their own actions (and lack thereof), they got a thrill out of seeing themselves on the monitor. It's interesting to note, also, that the *reduction in unit man-hours* (the man-hours required to place 1 cubic yard of concrete) was not achieved by reducing manpower but, rather, by increasing production.

By watching themselves on tape, the craftsmen were able to spot things they could have been doing during periods when their regular work had been delayed. Detailed assignments were traded back and forth amongst the craftsmen, and the end result was that the crew could place significantly more cubic yardage in a day than before, with no additional application of resources.

They realized that if they could reduce costs by getting the work installed more quickly, they might work themselves out of a job a little earlier in the short run, but, in the long run, their employer was going to be more successful in getting new work and thus keeping them working more continuously. This made their particular situations more secure in their own minds, and time proved that they were right.

In this case, before-and-after tapes were used in subsequent training classes not only to show the operations that each member of the concreting crew was to perform, but also to illustrate the manager's policy of encouraging full participation by craftsmen in determining improvements in their own methods and techniques. The original crew was proud to have been selected to star in the videotape sequence, and they derived a great deal of satisfaction from their participation.

12

Recognition of Achievement

Achievement brings self-satisfaction and the esteem of others, which, in turn, motivates the person to achieve again. The process repeats itself. It is human nature to like one's self, and when a person does something he is satisfied with, he likes himself even more. When he completes a task he has set out to do and is satisfied with what he has done, he recognizes his accomplishment and feels good about it.

SATISFACTION AND ESTEEM

The self-satisfaction that comes with achievement is a rich experience, and it induces the person who experiences it to seek out other goals, other accomplishments, and other satisfiers. It is tied closely to another satisfier, the increased self-confidence that comes from the accomplishment of something new. The two together create a desire within the craftsman to achieve even higher goals because of the expectation of more ego satisfaction and increased self-confidence.

More deeply satisfying to many craftsmen than self-satisfaction is the esteem of others that comes with the recognition of his accomplishments. Depending on his own personal values, the craftsman may enjoy recognition and approval by his supervisors, his fellow craftsmen, his subordinates, or all of them. He may get the greatest enjoyment out of recognition by his family and loved ones.

The nature of his accomplishment may also dictate the source from which he gets most satisfaction. For example, if he completes a training course in pipe welding, he may be motivated most by the esteem of his peers on the job site—those who knew him before as well as after the training. On the other hand, if he completes a course that is more academically oriented, such as a supervisory development course or a high school general equivalency diploma course, he may get most satisfaction out of recognition from his wife and children, for whom he might be acting as a role model.

In addition, the craftsman usually obtains satisfaction from the contribution he makes every day to the advancement of his project. Even though his labor may represent only a tenth of a percent of the total, he identifies himself with the overall endeavor—again, he projects. It is personally fulfilling for him to stand back and view the results of his efforts. In combination with other team members, he has produced something tangible, something that he can say he helped construct.

RECOGNITION

It is even more gratifying when he can point out his accomplishment to those who are most important to him, whether it be work associates, friends, or family. Through them, he gains recognition for his labor, and this transforms itself in his mind to esteem, which is really what he seeks—someone to see what he has done and to think a little better of him for it.

A project manager came to this realization by accident on a project several years ago. Doing some paperwork at home one Sunday afternoon, he discovered that he needed a particular document from his office and drove to the project site to get it. The project site ran several hundred feet along a rural highway, and the manager was surprised when he arrived and saw several cars parked along the shoulder of the road. A number of people were looking through the fence at the construction. He thought there might have been a fire or something similar and stopped. When he got closer, he saw familiar faces and realized they were craftsmen who had brought their wives and children to show them the project they were helping build. The craftsmen were gesturing continuously, and the families were evidently fascinated with what they were being told.

An idea formed in the manager's mind. He thought if the craftsmen's families were interested enough to come out and spend time on the weekend looking at the job through the fence, they might enjoy even more the opportunity to see it close up. He developed a plan to put on a "Family Day," a day on which everyone involved in the project—engineers, secretaries, accountants, and other staff members, as well as craftsmen—could bring their families onto the project and receive a guided tour of all the facilities, not just the portions that could be seen from the highway. Invitations were mailed out, and, on the selected date, the project manager and several supervisors conducted the visitors around the site in small groups. The tour guides were well-prepared and, in addition to describing the details of the project, took the time to point out where many of the craftsmen were assigned and which crews had performed which operations. The effort provided a double benefit. First, the initial purpose was served: the craftsmen were able to show their families what they had done and were doing in terms of pouring concrete, erecting steel, and pulling wire. Second, the craftsmen themselves received a detailed briefing about the whole project: the products to be made there and what functions the various areas would serve. The experience brought into focus more clearly just how the efforts of each craftsman, each crew, and each craft group was integrated into the overall picture.

This form of recognition was very well-received by the employees and was repeated several times as the project progressed toward completion. Motivational levels increased significantly, and productivity at the end of the project was at an all-time high for that contractor.

PLANNING RECOGNITION

The manager who thoroughly plans the manner in which achievement is to be recognized will amplify many times the sense of satisfaction that craftsmen derive from their accomplishments. This sense of satisfaction will motivate them to make the accomplishment again, even better next time, or even try something more challenging.

For many craftsmen, the most meaningful recognition is recognition of their completion of a craft or supervisory training program. If the manager takes the time to prepare the setting in which this recognition is given, the significance of the achievement will be enhanced in the minds of the receivers. A few inexpensive arrangements and preparations are all that is required. For instance, if a special ceremony to award certificates of completion is conducted, the craftsman's memory of his achievement will be richer than it would be if certificates were simply handed out in plain envelopes at the end of the last class.

Moreover, if the ceremony is held at a time when members of the craftsman's family can be present, he will never forget it. If some high company official not normally located at the project site presents the awards, the craftsman will derive not only a sense of self-worth from having met and shaken the hand of the "big boss" but also a sense of "belonging" to the company.

Other enhancements to this type of recognition include articles in company or project newsletters, local and hometown press releases, and copies of photographs taken during the ceremony for the craftsman's scrapbook. Almost anything that is visible and enduring will serve to emphasize the recognition and provide motivational energy for years.

The importance of managing the recognition process properly cannot be overemphasized. Careful attention should be given to the proper forum or vehicle for presenting it, the timing of the presentation, and the level of publicity to be given to it. An inadvertent oversight can largely nullify the benefits that might otherwise be derived. And the fallout of a poorly managed episode can affect an entire job site in a matter of days or even hours. One project manager learned this the hard way.

In an attempt to motivate a project work force to improve its attendance record, he instituted a "best-crew" program for the crew each month with the lowest absentee rate. To ensure fairness and impartiality, he put selection of the crew in the hands of the timekeeper, whose records of attendance were used for payroll purposes and were accepted by everyone as accurate. Things went well for several months. Absenteeism statistics were improving. But one month, unknown to both the timekeeper and the project manager, the crew selected for recognition had been given word immediately before the lunch that they were to be laid off that very day at the end of the shift. Their portion of the work was winding down, and fewer craftsmen were required. Unfortunately, the project manager was unaware of this turn of events when he sat down to lunch with the craftsmen.

Needless to say, all the recognition in the world could not improve the disposition of the crew during the "best-crew" luncheon that day. The crew's reaction was as might have been expected: they were puzzled. If they had made such a significant achievement, why were they being laid off? The fact that they were the next-to-last crew in that craft to be furloughed (the remaining crew being the most productive on the project) made little difference to them. What had been intended as a motivating experience ended up a demotivating one. The only positive fallout was a recognition that better planning and communication were needed. The manager, to head off negative talk in the field, explained honestly that his assistants and he—i.e., management—had erred through poor communication among themselves and apologized for the oversight. Fortunately, after a period of fine tuning, the program carried forward again with good results to the end of the project.

POSITIVE REINFORCEMENT

Recognition and *praise* are terms sometimes used by psychologists to describe the concept of positive reinforcement. To laymen, the concept means that giving praise when a desired task is accomplished and withholding it when it is not accomplished is more productive as a motivator than giving no praise when the task is accomplished and open criticism when failure occurs. This approach—*positive reinforcement*—is the antithesis of the "cussin' " approach described earlier.

Construction craftsmen, like everyone else, thrive on recognition and praise. The project manager, seeking ways to improve the motivational climate on his project, should continuously be looking for new ways to give deserved praise and to recognize genuine accomplishment.

Advancement is one way in which a craftsman expects to be rewarded for significant achievement. The manager should empathize with his craftsmen and decide whether he, himself, would be satisfied with the level of recognition he plans to give them for their accomplishments.

Take, for example, the craftsman who has demonstrated skill, knowledge, and leadership potential and is considered by his supervisors ready to take over the supervision of a crew. Management decides to promote him. At this point in the decision-making process, the manager should pause and think about the best way to handle the recognition that goes along with a promotion. Should the manager just tell the craftsman one morning that he is now a foreman and add a little money to his paycheck, or should he announce the promotion at the weekly job or staff meeting, post it on the bulletin board, note it in the project or company newsletter, and write a personal note to the craftsman's wife telling her how proud the manager is of her husband? Of course, the latter approach to recognition is going to be an infinitely greater source of satisfaction to the craftsman and, in turn, create a much more motivated employee.

Simply making the new foreman's paycheck bigger without the words of praise and recognition to go along with it can actually cause resentment and rejection. Without any explanation, it may strike him that management needs more foremen and is grudgingly paying him more only because everyone in the new category gets that amount, not because he has done anything special. No one has told him he is being promoted because he has done a good job and deserves it. He feels that his effort has not really been recognized. The praise that would have confirmed the promotion is missing. The uncertainty created by the lack of recognition raises doubts in his mind. He thinks he has performed well, but since none of his supervisors has acknowledged it, the higher esteem he has sought eludes him. His self-respect diminishes. While he will probably do enough to keep his foreman status, he likely will not put forth the effort to gain another promotion. That is his way of rejecting management—"getting even"—for the way he feels management has rejected him by withholding the recognition he feels he deserves for his accomplishment. What should be a motivator becomes a serious demotivator.

The cost to the manager of the difference in approach can be measured in terms of his time and effort, not dollars. It costs virtually nothing in dollars to emphasize the craftsman's accomplishment. What it does require is some effort on the part of the manager. He must write the note for the bulletin board, prepare a short statement for the staff meeting, and have a one-page letter typed up for the craftsman's family. These tasks might consume an hour of the manager's time—not much in contrast to the increased productivity the new foreman can achieve through increased motivation.

TIME AND PLACE

Other factors surrounding recognition should also be considered. If the decision to promote a craftsman is likely to be popular with his immediate associates—the crew he is going to supervise—then the manager should tell them before he tells the rest of the work force. Let them be the first to congratulate him and show their esteem for one of their fellow workers. If, on the other hand, the new foreman taking on a crew of strangers or new hires, then telling a wider group on the job would be better.

It is important that the news be a positive experience for the craftsmen—not a negative or even a neutral one. Obviously, if the new foreman's crew members have never seen him before, then their enthusiasm for his promotion will not be as great as it would if they had been working together for some period of time.

RECORD KEEPING AND CERTIFICATION AS RECOGNITION

In an organization within which craftsmen regularly move from job to job, it is important that up-to-date personnel records be kept on a central basis. All the motivational impact of an accomplishment on one job—such as becoming certified as a pipe welder—will be destroyed if, when that person moves on to the next project, his hard-earned status (and that's what it is to him) is lost because the employer has kept no central record or other means by which the craftsman's association with the organization can be tracked by personnel administrators and supervisors.

It is analogous to a fourth-grade pupil transferring from one school to another within the same school district and being put in first grade because no central record of his having completed the first, second, and third grades exist.

Fortunately, with desk-top computers proliferating everywhere, and with many of them networked into mainframes in corporate home offices, more and more employers are keeping better records on their people so that this loss-of-recognition problem is easier to prevent.

To assure that records are accurate and as current regarding a craftsman's qualifications as possible, some construction companies have instituted craft certification programs as adjuncts to their training programs. Patterned after welder certification programs, which have been in existence for several years, craft certification programs allow continuous tracking of an individual's performance with a firm. His demonstrated skill levels and other career development information are stored so that at each engagement with that firm, he is assured of being hired at the proper level.

To obtain certification at a certain level in a craft, the craftsman must periodically pass a written test to confirm his knowledge and a functional test to demonstrate his skills. The tests are prepared by the company's most knowledgeable employees in the particular craft and are administered under strictly controlled conditions to ensure fairness and impartiality. As improvements in construction technology occur, tests are modified so that recertifications will be up-to-date.

Through accurate personnel record keeping, the employer can keep tabs on his good employees for future assignments. This not only benefits the employee because his accomplishment is recorded permanently, but also provides him with a measure of recognition. Just the fact that his name is on a list somewhere back at the home office serves as a motivator for the craftsman. First, he feels that he is, to some extent, part of an organization (everyone likes to feel they're part of something). Second, he is pleased that the employer cares enough about him to want to keep a record of his achievements and progress in the company in addition to the data required by the government.

VISIBLE EVIDENCE

Of course, it is even more of a motivator if the craftsman is given some permanent record to keep for himself. Again, it takes little effort on the part of the manager to prepare appropriate award or promotion certificates and give them to the craftsman so that he, too, may have some permanent record of his accomplishment. Hanging on a wall at home, where it can be seen every day, a certificate is not only a constant reminder to the craftsman of the standards to which he holds himself and of what he has accomplished but it also provides pride and self-confidence. It is visible encouragement that motivates him to continue excelling.

An industrial psychologist studying the relationship between achievement and visible evidence once performed an experiment at a bowling alley. He asked a team of bowlers to explain what it was that they most enjoyed about the game. They replied that they got the most satisfaction out of rolling a high score. He then proceeded with the experiment. The first bowler lined up, made his approach, and executed a classically beautiful release. As soon as the ball was away, the psychologist signaled an assistant who caused an opaque curtain to drop over the alley, obstructing the bowler's view of the pins.

The surprised bowler heard the ball hit the pins and saw the result–a strike–appear on the scoreboard, but he complained vigorously that he couldn't see the pins fall. The psychologist reminded the bowler that he had said he obtained his enjoyment from his score and therefore missing the sight of the pins falling shouldn't have disappointed him. The bowler then acknowledged that the score, by itself, was not enough; he needed to see the physical result of his effort. To gain satisfaction out of the game, he needed the visible evidence of seeing the pins fall down.

So it is with the construction craftsman. The bowler's score is analogous to the craftsman's paycheck. The sight of the pins falling is analogous to the recognition the craftsman seeks when he does something well on the job. The craftsman needs something to look at that will remind him of his achievement, whether that something is a certificate hanging on a wall at home, a special recognition decal for his hard hat, or a short personal letter from the boss thanking him for some extra accomplishment.

SPECIAL RECOGNITION

Recognizing achievement by granting special privileges when appropriate is another powerful motivation. This type of recognition is usually best reserved for situations in which a competition of some sort results in a winner or a small group of winners every month. For example, there might be a "craftsman-of-the-month" award based on safety, attendance, productivity, or workmanship. The winner might be given a special-color hard hat with his name, the achievement, and the date hand-lettered on it. Anything that might be coveted by most craftsmen can be used as the trophy, but it should be attractive and highly visible so that everyone can associate the achievement with the achiever.

Another example of special recognition might be a reserved-parking spot for the winner for a period of time, typically a month. Some location that is considered highly desirable by everyone in the work force should be selected. In some cases, it might be the slot nearest the entrance gate to the project; in others, it might be the one closest to the exit; in still others, it might be the one that is grass instead of dirt or the one that faces away from the afternoon sun. The key is to find a prize that most craftsmen would like to possess enough to compete for it.

The craftsman must be interested enough in the prize to try for it. But also, the recognition he contemplates getting when he wins must be fully what he expects it to be. If not, the word will get around that the prize is not worth the effort, and the program will lose its effectiveness.

Other types of recognition take on more continuing forms. Included in these are the traditional perquisites that accompany responsibility. Privileges related to level of responsibility are present in a free enterprise society because they represent achievement and success. For a worker, they are tangible symbols of the status he has achieved when he advances into the supervisory ranks. Some of these symbols, such as white hard hats for supervisors versus colored ones for workers, are simple—but coveted. Others are related to convenience and comfort.

It is necessary to draw a clear distinction between conveniences that are, by today's standards, basic needs and those that go beyond the basic. Everyone must get their basic needs tended to; anything beyond these needs can be recognition for achievement. Figure 12-1 illustrates the difference between basic needs and perquisites by hierarchy level. Everyone in the organization is entitled to basic needs, points a_1 and b_1, while those closer to the top may also receive perquisites, points a_2 and b_2. These fall into several different categories, starting with working conditions.

At the basic-needs level, all employees should have access to clean toilet and washroom facilities. The personal dignity of the employee requires this. Also, all employees should have a clean place, protected from the weather, in which to eat their lunch and ample cool drinking water at all work locations. These are examples of fundamental personal needs that the employer should furnish to employees at every hierarchical level on the project.

Taking the eating-area example further, craft superintendents often have field sheds or offices in which they eat their lunch.

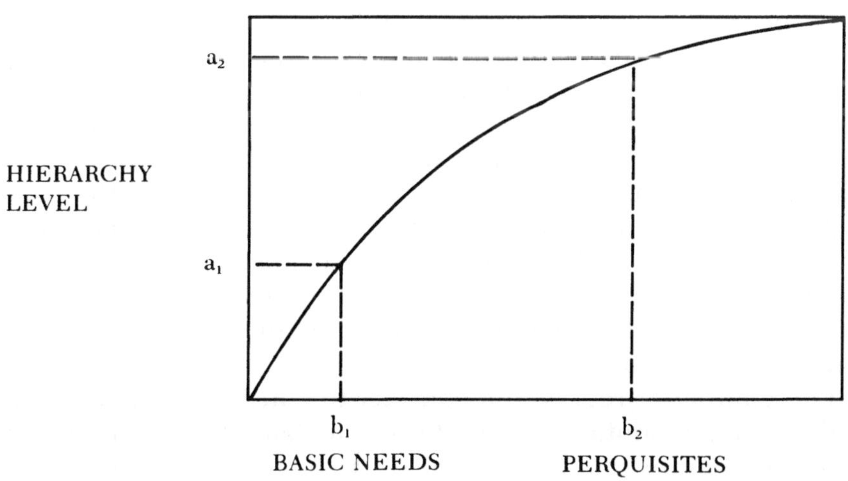

Figure 12-1 Hierarchy Level vs. Basic Needs and Perquisites

If the sheds were air-conditioned, this would represent a perquisite. The perquisite might be a form of recognition of the fact that the superintendent has progressed up through the organization to his current level.

Similarly, an employer might permit certain supervisors to use a company-owned vehicle for their personal needs as a form of recognition for special accomplishment. This is one of the more coveted perquisites in the supervisory ranks and one that should be carefully controlled by the manager because of both its impact and its cost. As such, it is one above most others that should be reserved for recognition of outstanding achievement.

LOSING DISTINCTION

The concept of perquisites can be abused if not managed properly. To be successful, the perquisite should be carefully maintained as recognition for achievement. It should not be indiscriminately awarded. If it is, its basic motivational benefit—its distinctive value—will be lost; in some cases, even more serious consequences occur.

One contractor lost control of the use of his pickup trucks. For many years, he allowed supervisors whose jobs required extensive travel to keep their trucks full-time, on the job and off. It was a form of recognition for the extra effort and the long hours they spent on the road. However, over a period of several years, the distinction became blurred, and, eventually, virtually every supervisor had the full-time use of a pickup truck whether he needed it or not. What started out as a perquisite became, essentially, a custom. It became a condition of employment and was considered as coming with the job; i.e., as part of the compensation to which all supervisors were entitled. It thus lost its value as recognition for achievement. In addition to costing the contractor a substantial amount of money each year for fuel and maintenance, without any real benefit to him, it eventually cost him a valuable contract.

The contractor received a contract to construct a large mine high up on a mountain in a western state. Safety rules prohibited gasoline engines inside the mine, so every day, each of twenty-two foremen drove his truck from his quarters 9 miles away up the mountain and parked it for eight hours while he worked inside the mine. No particular individual or group achievement was involved, and thus no special recognition was required. Even though it would have been far less expensive and a lot safer to take the foremen up the mountain to the project site in a bus, each supervisor drove his own truck; i.e., exercised his "right."

Eventually, the sight of those twenty-two trucks parked along a widened-out spot on the narrow haul road, which had been blasted out of the side of the mountain at a cost of over ten dollars per square foot, was enough to convince the owner, who was paying all the bills, to terminate the contractor in favor of another with somewhat more pragmatic personnel and recognition policies.

REASONABLENESS AND CREDIBILITY

Judgment is called for in all situations, and the quality and quantity of recognition accorded any individual or group must be analyzed for its reasonableness.

There are two basic requirements when it comes to recognition of achievement. The first is that the recognition be offered by the giver and perceived by the receiver as genuine and well-earned. The second is that the process of selecting who is to receive recognition, and under what conditions, be perceived by everyone affected by it as honest and impartial.

A bogus award, one given for some sham accomplishment or one whose magnitude is far out of proportion to the goal achieved, will be a demotivator and cause the giver to lose credibility.

The craftsman receiving a bogus award knows that it is not deserved. Those whose esteem he seeks, his peers and superiors, recognize the same thing. The award is deemed meritless by these groups, and they look for ulterior motives behind the giver's actions. Suspicion develops, and credibility suffers.

The second basic requirement, that the selection process be based on honest and impartial evaluation, means that the craftsman must, first, understand the rules under which the contest is being conducted and, second, must perceive that the contest is being carried out fairly. The slightest hint of partiality will damage the program on that particular project beyond repair and may even cause an irreparable credibility problem in the company itself.

The manager must therefore be extremely cautious in designing and implementing any recognition program. While a properly executed program can reap an abundant harvest in terms of motivation and productivity, a poorly implemented one can cause damage far out of proportion to any potential gain.

GROUP RECOGNITION

Recognition can be for many different accomplishments and can be oriented toward an individual or toward a group. Group recognition has the added benefit of promoting team pride. A group of craftsmen cooperating with each other to reach a specific goal produces a sense of comradeship that can extend into everything they do as a team. They learn to anticipate each others' needs, actions, and reactions; as a result, their activities become better-coordinated. Improved work flow is achieved, and productivity improves.

DIRECT PRODUCTIVITY TRAPS

Setting up a system of direct recognition for units produced per man-hour requires careful planning and execution by the manager. Setting up a competition between crews to see which can outwork the other may be perceived as exploitation and thus quickly dismissed by the craftsmen as a management ploy. In addition, there are other traps that may cause more overall damage to the project than this type of recognition can produce. For example, if improved productivity is gained at the expense of quality of installation or safety, the end result could be higher project cost due to reduced motivational levels among the entire work force. Good craftsmen take pride in the fruits of their labor and do not get any satisfaction out of doing a poor or incomplete job.

Moreover, when the contest is over, what then? If the project continues, there will be a slackening off in units produced. If the project is at completion, the same type of contest will have to be conducted on the next project or output will similarly suffer. Eventually, all work involving that group of craftsmen will be subject to the same requirements. It will become routine to expect some form of competition to achieve ordinary production goals, and the recognition value will thus be lost.

The choice of parameters to determine achievement goals can make a critical difference. The manager who blunders into a competition based solely on units produced per man-hour will encounter the problems described above. But the one who uses parameters related to output, but less directly, may easily accomplish the same goal. For instance, if the competition were, instead, based on the improvement-in-performance factor—the ratio of estimated to actual productivity—and a graphic representation of the results were shown to all contestants weekly, the same improvement in productivity could be accomplished but without exploitive overtones. The contestants would be focused on surpassing a standard such as the job estimate or their past performance. A crew with a poor performance factor might take up the challenge just to get back to average—"out of the cellar," so to speak. In addition, the award of worthwhile prizes to the winners serves to enhance the recognition value as well as stimulate the competitive aspects of the contest. Graphing of performance factors is illustrated in the cost-control discussion in Chapter 13.

RECOGNITION PARAMETERS

For all but the largest projects, those staffed with professional cost engineers and technicians trained to administrate cost control and work sampling programs, it is more practical to set up competition for recognition in activities such as reduction of accidents, personnel turnover, and absenteeism.

For instance, competition among all the crews on a project may be developed as a means of improving the project accident-prevention record. The best crew of the month is selected as described in Chapter 5 and awarded safety medallions for their hard hats. The crew is given further management recognition through a private luncheon with the project manager. It is a win-win situation, for the craftsmen get the praise, self-satisfaction, and esteem of their peers for having done something special while management gets to talk to the craftsmen about job conditions and problems and gets to put emphasis on accident reduction. Both of these benefits lead to better productivity on the project. The recognition the craftsman receives creates ego satisfaction and the desire to do better in all regards, including the category in which the award was bestowed. At the same time, the manager can use the value of the information gained to make improvements and reduce costs.

Sometimes, good-natured, fun-type contests can be conducted with spectacular results. The key is to find something the craftsmen enjoy doing, talking about, and winning. One manager solved a troublesome safety-awareness problem with just such a campaign. On a particular hazardous plant expansion project, the safety record was not nearly so good as it should have been. Accidents were occurring because the noise, fumes, congestion, and confusion that accompany work in an existing operating facility were affecting the workers' concentration.

In order to focus attention on the need for craftsmen to think about safety all the time, the manager devised a contest in which the winning crew would be given a cash bonus. Each crew was assigned a 16-foot-tall post that was planted in the ground next to the entrance gate to the project. The size of the award to the winning crew would be determined by the length of their post at the end of the contest.

Each week, the safety engineer tallied the number of unsafe acts and conditions he observed for each crew. At lunchtime on the last day of the week, with great ceremony, the safety engineer cut off a portion of each crew's post according to its safety record for the week. An inch was cut off for each infraction of the safety rules. At the end of the competition, each member of the crew with the tallest post was awarded a certain amount of cash for each inch remaining on the crew's post.

It cost the manager several hundred dollars to pay off the winning crew, but, in the meantime, lost time on the project due to first aid cases and visits to the doctor and medical facilities dropped significantly, with a commensurate savings. The safety-conscious attitude that had been rejuvenated during the contest survived, and the safety record on that project ended up being one of the best in the company.

13

MEASUREMENT OF RESULTS

The result of a successful motivation program is improved productivity. It can be evaluated analytically and subjectively in several ways. Analytically, it can be directly measured through the management information system reports that are generated in most companies and through mandatory federal or state agency reports. Study of this information will reveal a trend toward an increase in productivity and a decrease in cost parameters when compared with a project without a program.

It can be evaluated subjectively by looking at more subtle indicators. This evaluation requires that the manager empathize with and be cognizant of the actions and reactions of craftsmen. He will detect from analysis of these that they are in a relatively happy state of mind.

ANALYTICAL EVALUATION

Cost-Accounting System

The primary analytical indicator on a project is the long-range trend in labor productivity. Many companies use a basic cost accounting and reporting system in which major blocks of labor cost—e.g., the cost of all foundations or of all electrical work—are compared periodically with an estimate that was prepared before the work began. The comparison shows, at any stage of completion, the difference between estimated and actual costs. If actual costs are less than estimated costs and the difference between them increases over the course of the project, then productivity is evidently better than anticipated, provided no changes have taken place since the estimate was prepared.

This elementary system, while suitable for small, short-duration projects, has several drawbacks when stretched to fit a larger one. Some of these may cloud the true performance of labor in the field. First, the basic system usually does not go into enough detail to permit control and projection of costs, only reporting of what has occurred in the past. It precludes factors such as the efficacy of planning, the quality of supervision, and other support functions from being evaluated as causes in the event of adverse cost departures. Second, such a system is not versatile enough to accommodate changes in quantities effectively. Quantity variance affects both labor and material costs directly, and without accurate tracking of changes, no meaningful control can be exercised. Third, any analysis made before the project is finished must, of necessity, reflect the status of physical progress toward completion. Much of the time, determination of this critical parameter is left to superintendents, who usually overestimate progress; to accountants, who usually underestimate it; or to untrained personnel, who may be inconsistent, overestimating progress one time and underestimating the next.

One contractor, disappointed at the wide swings in unit costs he saw in his job-cost reports, found that it was due, in large part, to the manner in which percentage complete was determined. An accountant had been assigned the task of evaluating completion status. He reasoned that if, for example, 30 percent of the budget money had been spent, the project was 30 percent complete. He made no allowance for quantities in place or any other installation-related influences. Consistency of the contractor's cost units became more reasonable after he placed determination of completion status in the hands of more knowledgeable personnel, who based their calculations on actual quantities in place compared to total projected quantities (including scope changes).

Cost-Control System

Most companies working on larger projects watch trends in labor productivity very closely. They use some form of cost-control (in contrast to cost-accounting) system in which the units of labor—usually man-hours—required to produce a given unit of work, such as cubic yards of concrete or linear feet of pipe, are calculated each week. This system combines a work-account-coded timekeeping system with physical quantity surveying and produces unit costs for all activity accounts on the project.

The total number of man-hours charged by all foremen on their crew time cards to the account number, or "cost code," for footing concrete placement, for example, is determined by the timekeeper when he makes up the payroll for the week. A cost engineer or quantity surveyor measures the total cubic yardage of concrete placed in footings that week and derives a *unit man-hour cost*—the man-hours expended to produce one unit (i.e., one cubic yard of concrete). Comparison with the unit cost for footing concrete used in the estimate, sometimes called the "bogie," indicates relative productivity.

When the ratio of estimated unit cost for a work item to actual unit cost is calculated, a direct measure, called the *performance factor,* is determined. If the performance factor is greater than unity (1.0), the work has cost less than was projected in the estimate. If it is less than unity, the work has cost more than the estimated amount. The cost engineer produces graphics to show weekly and cumulative values. Trends in productivity are easily spotted through performance-factor graphing over time. The positive slope of a trend line indicates improving productivity. Figure 13-1 shows a typical performance-factor curve over the course of a project.

This type of system allows close control of the project because the feedback on the effort expended by the work force is available every time the payroll is closed. Unfavorable trends can be investigated immediately and remedial action can be taken. The action may consist of rescheduling the craftsmen in a particular area to reduce congestion, adding hoisting equipment temporarily to a project task, or making a similar correction.

Craftsmen welcome this type of project fine tuning because it allows them to do their jobs more efficiently. The benefit is thus magnified. The craftsmen are not only released to practice their skills and produce more but are also, because of the signal they receive—i.e., support from management—motivated to strive for higher levels of productivity.

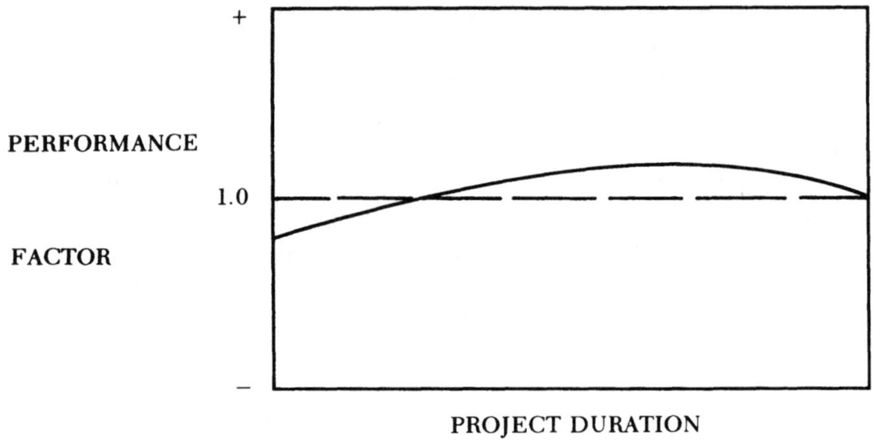

Figure 13-1 Performance-Factor Curve

Work Sampling

Another direct indicator used by some companies is called *work sampling*. It can be used as the primary indicator of productivity or in support of either cost-reporting/control system described above. Rather than focusing on units of work produced per manhour, it looks at what craftsmen are doing at several randomly selected times during the week.

In this system, each crew is assigned a discrete identification number. "C4," for example, might identify the fourth carpenter crew formed on the project. Each member of the crew is issued a hard hat on which the identification number C4 is clearly visible. A trained observer walks around the entire project at frequent intervals and determines the type of activity each craftsman is engaged in when observed.

Observations are grouped weekly by activity, and a statistical profile for each crew is developed. A report is produced that indicates the percentage of observations, and therefore the time, each crew spent on three main categories: (1) their basic job, (2) auxiliary work and (3) delays. It shows, in detail, the percentage of time crew members were observed performing their basic tasks, such as sawing lumber, bolting up steel or laying down a weld bead. It also shows how much time they spent performing auxiliary or support work, such as traveling to procure materials or tools, and how much time they spent waiting for materials, standing in line, waiting for other craftsmen to finish a preceding task or engaging in similar nonproductive activity.

On a project with a successful motivation program, the percentage of time attributable to delays will be lower because the work has been better planned. In many cases, craftsman input, such as the concrete crew described in Chapter 11 who used time lapse photography to improve their productivity, is the major ingredient in better planning. Also, the percentage of time spent traveling will be lower because management has arranged the location of toolrooms and supply points so as to minimize the need for craftsmen to travel far to pick up an item.

The percentage of time spent on the second category, *auxiliary work*, will also be lower. The motivated craftsman is thinking ahead and planning his own activities to minimize the preparation time required. When he goes to the supply room, he gets everything he needs, not just a portion, for the immediate operation at hand. Figure 13-2 shows the percentage of time spent on the three main categories on a well-managed project.

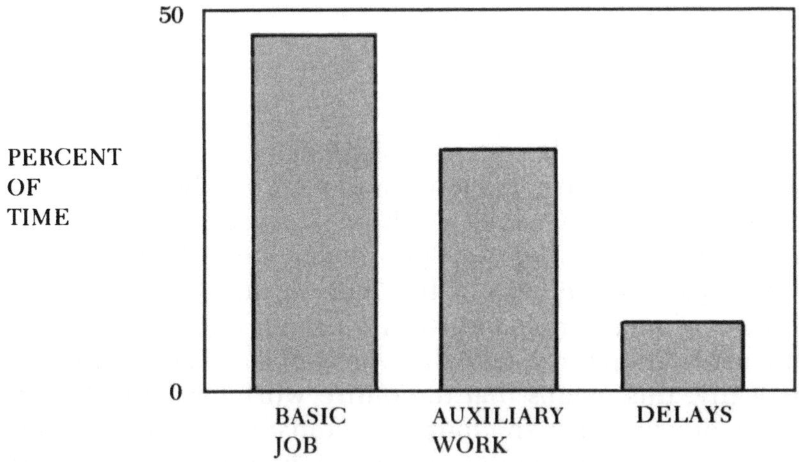

Figure 13-2 Work-Sampling Bar Chart

Companies that use work sampling keep long-term statistical data from which to set baselines for any particular type of project in the geographical areas in which they operate. This information is one more way to gauge the efficacy of a motivation program.

Absenteeism

Other statistical data also point out the improvement in job operation due to a motivation program. A project with a sound program will experience lower absentee rates than a project without one. If daily absenteeism on a mediocre project runs, for example, 8 percent, a project with a motivation program in place may see only a 3 or 4 percent rate. This is because on the poorly run job, it is sometimes an emotional struggle for craftsmen to come to work each day. They know there will be confusion and conflict, congestion and delay when they get there. They know they will have little opportunity to practice their crafts without a lot of frustration. Some, therefore, give in to the struggle and stay away every so often. In contrast, on the well- run project, craftsmen are usually there not only every day but, more important in confirming their motivational level, early. They want to get to work; they look forward to things going as planned and to the opportunity to derive satisfaction from performing the tasks for which they were trained.

Turnover

Similarly, *turnover*, the number of times per month a specific job slot must be filled with a new employee, will be considerably lower on a project in which a robust motivational climate has been created. Typically, on a poorly run project, this statistic will be 15 percent, perhaps higher. This means that the manager on a 100-craftsman project must hire fifteen new craftsmen per month just to keep the size of the work force at its original level. Statistically, this means that the entire work force must be replaced every six to seven months. A craftsman sometimes does not become fully productive for as much as three weeks after he has started work on a new project. This is due to the unfamiliar surroundings and the time consumed by the hiring-in and orientation procedures necessary today. The new employee needs time to learn where to go to get tools, materials, and equipment and even to find the toilets and washrooms. In addition, it takes time to become familiar with working with a new crew and for a new foreman.

Craftsmen on a project with a strong motivational atmosphere do not experience the nagging frustrations that cause them to leave. They like the environment and want to stay as long as possible. This means that the manager will, in addition to lower turnover rates, experience significantly lower labor costs.

Safety

Still more analytical evidence of the level of motivation on a project is found in the reduction of accidents. Federal regulations require that detailed records be kept on first aid cases, doctors' cases, and lost workday accidents. Assuming these data are reliable indicators of the level of unsafe conditions and unsafe acts on a project, the project with a high motivational atmosphere will have a significantly lower incidence of accidents. When workmen are not consumed with frustration and not spending a lot of thinking time about work-flow problems and their own dissatisfaction, they can keep their minds focused on the proper safety procedures and be more alert to potentially hazardous conditions. They do not want to be even slightly injured, much less suffer an accident that will keep them away from work, because work is a pleasurable and rewarding experience for them.

Quality

Though not as readily measurable as some of the above indicators, the quality of installation on a project with a motivated work force will be higher. The cost of remedying defective work can be isolated and kept for cost purposes. However, in a positive sense, the reduced cost of getting quality craftsmanship the first time an operation is performed is usually only detectable in the final cost units and performance factors. Subjectively, however, the presence of quality workmanship will be instantly recognizable to an experienced superintendent.

SUBJECTIVE EVALUATION

A more subjective measure of the success of a motivation program is the readily perceivable transition of the job site to vibrant, dynamic activity. A certain hustle and alertness will be readily apparent in the work force. Craftsmen will feel, look, and act action-oriented. The desire to excel will be plainly visible up, down, and across the organization. Grumbling will shrink to normal, healthy levels, and other signs of dissatisfaction, such as bathroom graffiti, will diminish or disappear.

Beating the Estimate

For the manager, the best measure of a successful motivation program will be the productivity evidenced in the final cost of the project. In an improving motivational atmosphere, this cost will be less than initial projections. The initial estimate is usually prepared from historical data, and the higher performance factors that result from a motivation-productivity improvement program will not be reflected in the units used for estimating the project. Their influence will show up on the next project, when lower unit costs can be substituted for previously higher ones. Obviously, if the program continues, the next project will also beat its estimate. In any case, a contractor with a motivated work force can compete more successfully than one without.

Labor Relations

Relations between labor and management will show appreciable improvement on a project in which a healthy motivational climate has been developed. These results are also less analytical in nature insofar as assigning a dollar value is concerned, but they are nevertheless visible and sometimes more dramatic to observe.

In an open-shop environment, signs illustrating a healthy relationship between labor and management are somewhat more subtle than they are in a union environment because the absence of certain negatives is apparent. What will be noticed is a reduced level of griping both on the job and off. Craftsmen tend to congregate after work at one or more central locations to socialize and enjoy a cool refreshment. A manager listening to the gist of the conversations at these places will notice that the talk is not about poor conditions on the job. He will hear that the craftsmen think things are going well and that they enjoy their work. The griping, if any, will be about the failure of one of their fellow craftsmen to do something properly. A form of self-policing actually takes place in a motivated work force, and the errant craftsman will be persuaded by peer pressure to correct his shortcomings.

The reduction in gripe level will also be accompanied by an ancillary benefit, a reduction in the appeal of the labor union. Historically, trade union organizations have tended to attract workmen who were upset, frustrated, and dissatisfied with working conditions on their jobs. When the craftsman is satisfied with the conditions of his employment, there is little griping and, therefore, no reason for him to lean toward a union. He reasons that there is not much a union can do for him.

Projection

Still another subjective measure of the efficacy of a motivation program is the degree to which craftsmen identify with the company. This is another form of projection, discussed earlier. In this instance, it confirms that the craftsmen respect management and have a positive image of it.

The craftsman wants to be perceived as being associated with the company and to be viewed as a contributor in some way to its success. More importantly, he wishes to be viewed as having the same fine character as the company and the other people within it. He thus mentally projects himself into the image of the company. He hopes and believes that when someone is impressed by something the company has done or is doing, he will be equally impressed with him.

One visible manifestation of this projection phenomenon is in wearing apparel and advertising novelties. The craftsman may wear a jacket or baseball cap with his company's name on it.

He may put a company decal on his lunch box or carry a lighter or pen-and-pencil set with the company logo on it. Anything that tells the observer there is a connection between the craftsman and the company will be sufficient. If the company makes these items available, either through awards or through purchase at modest prices, the manager will be able to get a good indication of motivational level on the job simply by observing baseball caps and jackets when craftsmen are entering the project or, even more significant, when they are in shopping malls with their families.

14

PROGRAM OUTLINE

Implementing a program to improve motivation and productivity in a company or on a construction project requires a considerable amount of front-end effort. It also requires that management focus its attention on the how-to aspects rather than the more philosophical aspects.

MANAGEMENT COMMITMENT

Further, if it wants such a program to be successful, management must do more than give it lip service. It must be sold from the top down. And the crucial test of management's dedication is the amount of time and money it commits to the program. If it is willing to allow staff members to carry out their program assignments, as either additional or full-time duties, and to commit funds to cover their time and expenses, then the program will get off to a good start.

If management continues to support the program throughout its early years, the bottom-line improvement will be evidence enough to convince even the most vocal critics of the program's efficacy.

INITIATION

The initial implementation step must come from top management. This is done by preparing a policy statement briefly confirming to everyone in the organization that the chief operating officer is serious about improving productivity by improving motivation. He is going to back the development and execution of a program designed to bring this about. He expects line and staff managers to learn about the program and to modify their styles, if necessary, to support it. He sets out definitive goals so that each manager knows his contribution to the success of the program will be evaluated when his job performance is reviewed. He also makes sure that recruiting and hiring representatives will seek out management and supervisory people with personality profiles compatible with the program.

PROGRAM OUTLINE

TOP MANAGEMENT ACTIONS

- Issue policy statement.
  - Confirm commitment of top management.
  - Support development and execution of specially designed program.
  - Counsel all managers to support program.
- Set out definitive goals.
  - Identify specific subgoals for each manager.
  - Relate an individual's contribution to his performance evaluation.
- Recruit supervisory personnel with motivational potential.
  - Seek out candidates for new and replacement job positions who have personalities compatible with motivational-management styles.
- Select Program Manager.

PROGRAM MANAGER ACTIONS

- Design motivation-program organization using team approach, with captains for each major functional responsibility.
- Identify team captains and assign to functions.
  - This responsibility should be added on to other job activities.
  - Avoid conflicts of interest.
- Prepare and present briefing material on motivation and productivity.
  - Reiterate top-management policy statement.
  - Prepare additional policy statements regarding the following:

      the dignity of the individual
      the consistent application of policies and procedures

- Introduce concepts.
  - Develop functional organization chart. Explain roles of each manager, supervisor, and staff employee in support of craftsmen.
  - Explain McGregor theories.
  - Explain Maslow and Expectancy theories.
  - Explain the eight essentials to the point of the work.
  - Emphasize
    - the need for correct sequence
    - the need for correct timing
  - Illustrate
    - direct support
    - direct communication
    - consistent management
    - sources of frustration
    - fine tuning the work force
    - assuring a safe workplace
    - basic personal comforts
    - training
    - recognition of achievement

TEAM-CAPTAIN ACTIVITIES

- Develop checklist of possible symptoms to evaluate.
- Obtain data in assigned area of responsibility by
  - personal observation
  - talks with craftsmen
- Consolidate and compare data obtained.
- Analyze symptons for root causes.
- Prepare detailed action plans.
  - Define specific actions to be taken.
  - Estimate cost to complete actions.
    - Do not assess cost against elements that are required by law, regulations, or preexisting company policy.
- Develop summary implementation schedule.

PROGRAM DESIGN AND APPROVAL

- Program manager evaluates detailed action plans.
  - Eliminates duplication.
  - Coordinates cross-responsibility plans.

  - Prepares program estimate.
  - Prepares milestone schedule.
- Set definitive program goals based on past performance for the following:
  - labor-performance factor or unit-cost improvement
  - workmen's compensation loss ratio improvement
  - monthly turnover ratio decrease
  - daily absentee ratio decrease
  - rework percentage decrease
  - QC (quality control) nonconformance incidence decrease
  - on-time activity completion increase
- Present program to management.
  - Obtain explicit management approval.
  - Obtain funding and timing commitment.

IMPLEMENTATION

- Set up program controls on the following:
  - cost
  - schedule
  - reporting
- Implement detailed action plans.

CHECKLISTS

Each team captain should prepare a checklist of symptoms to look for in his area of responsibility. A typical checklist might be made up of some or all of the following questions:

SUPERVISION

- Are supervisors aware of craft productivity levels?
- Are supervisors aware of project absentee and turnover levels?
- Are supervisors aware of project safety statistics?
- Are supervisors aware of project rework percentages?
- Do any supervisors and managers use the "cussin'" approach?
- Do supervisors and managers empathize with craftsmen?
- Do supervisors recognize that part of their role is to act as the craftsmen's advocate?

- Do supervisors have appropriate craft skills?
- Have supervisors had supervisory training?
- Are supervisors sometimes overbearing and oppressive?
- Does an elementary school attitude prevail?
- Does management evaluate the impact of proposed changes on craftsmen?
- Do some noncraft personnel see their role as the most important one on the project?
- Do some departments see their function as an end in itself?
- Do some personnel fail to see that their role is to support craftsmen?
- Are craftsmen properly instructed regarding the use of methods, procedures, tools, and equipment?
- Do craftsmen understand supervisors' instructions?
- Do foremen help plan out work schedules?
- Do craftsmen have an input into work plans and schedules?
- Are large increases and decreases in manpower level in the same craft frequent?

DIRECT SUPPORT

- Do all materials for a given task arrive at the point of the work prior to need?
- Are required tests and inspections performed without delaying craft crews?
- Is engineering information clear and complete?
- Is everyone involved committed to a common project schedule?
- Is schedule accomplishment monitored by management?
- Is schedule used to forecast and update manpower requirements?
- Is schedule fine tuned to level out peaks and valleys in required manpower levels?
- Are schedule-review meetings held regularly?
- Do foremen know what their schedule objectives are?
- Do craftsmen know the schedule objectives for their crew?
- Are materials and equipment to be installed delivered to the job site on schedule?

- Are materials and equipment warehoused in an orderly and easily retrievable manner?
- Is everyone informed about changes in plans or schedules?
- Is weekly labor-cost performance reported to management?
- Is weekly labor-cost performance revealed to each foreman?
- Are craftsmen required to travel far to pick up tools and supplies?
- Are sources of chronic delays identified and corrected?
- Are crews sized to fit tasks at hand?
- Are tasks broken into crew-sized components?
- Is crew mix modified to fit tasks?
- Are too many or two few craftsmen on the job for the work load at hand?

COMMUNICATION

- Are management's signals perceived as negative?
- Are management's signals incorrectly interpreted by craftsmen?
- Is management's credibility with craftsmen low?
- Does conflict between craftsmen and management frequently remain unresolved?
- Is an open-door policy in effect?
- Has management established effective two-way communication with craftsmen?
- Do craftsmen feel that communication with management is at a satisfactory level?
- Does the manager frequently "walk the project" to talk with craftsmen?
- Does the manager know craftsmen by name?
- Does the manager know the language of the trade?
- Do craftsmen feel comfortable with the manager?
- Are craftsmen relatively open with the manager?
- Are craftsmen given an employment orientation?
- Do craftsmen give suggestions?
- Do craftsmen voice complaints?
- Can craftsmen use a hot line to communicate with management?

- Are employee complaints and suggestions acknowledged by management?
- Do craftsmen perceive that management looks into complaints seriously?
- Are craftsmen's suggestions actively solicited?
- Are job rules published and visible to everyone?
- Are bulletin boards convenient, visible, and up-to-date?
- Are newsletters, personal notes, and paycheck inserts used to convey information?
- Are craftsmen informed about job schedules and progress toward the objectives?
- Do craftsmen know, consistently, what assignments to expect at least a day ahead of time?

CONSISTENT MANAGEMENT

- Do craftsmen feel that they are part of a team effort?
- Are job rules applied evenly to everyone on the project?
- Are visitors required to comply with job safety rules?
- Do office employees have rules and privileges more liberal or more desirable than craftsmen?
- Are absence rules applied consistently to everyone?
- Are humane exceptions to rules made for everyone?
- Are job rules frequently changed?
- Is any person or group perceived by others as "getting away" with something?
- Are there instances of "building a case" against an employee?
- Are supervisors frequently the first ones to leave the job site at quitting time?
- Are supervisors' perquisites perceived as out of proportion to their achievements?
- Are unskilled workers earning skilled-worker pay rates?
- Are there spurts of cleanup activity just before a VIP visit?
- Are basic personal comforts reduced or eliminated later in the project?
- Are group competitions perceived as fair and unbiased?

SOURCES OF FRUSTRATION

- Do crews frequently leave an area without completing all their tasks?

- Must a crew frequently wait for another crew to finish before beginning work?
- Is there frequently avoidable congestion at the point of the work?
- Are craftsmen frequently delayed when obtaining tools and supplies?
- Are tools frequently inoperable?
- Does construction equipment frequently arrive late at the point of the work?
- Does construction equipment frequently break down?
- Are required safety inspections performed expeditiously?
- Are unskilled workers given skilled-work assignments?
- Are there frequently errors in time, pay rates, or pay?
- Are paychecks frequently distributed later than scheduled?
- Are craftsmen frequently delayed when clocking on or off the project?
- Are tardiness and absenteeism levels high?
- Is the accident rate high?
- Do craftsmen feel abused or exploited?
- Do chronic personnel problems remain unresolved?
- Does management conduct exit interviews with quitting employees?

FINE TUNING THE WORK FORCE

- Are high levels of errors present?
- Are schedule elements and milestones frequently missed?
- Are job applicants screened for skills? Thoroughly?
- Are physically limited employees assigned strenuous work?
- Have any employees lost time on more than one project due to physical problems?
- Have any employees been repeatedly injured by temporarily disabling but nonserious accidents?
- Does an overview of the project indicate that many employees are constantly on the move?
- Are the same employees often observed carrying the same "props"?

- Do any employees tend to drag out casual conversations?
- Do any employees seem always to stop others from working just to talk?
- Do any employees seem to be constantly looking around rather than concentrating on their assignments?
- Do foremen know where their crew members are at all times, and can they point them out?
- Do any employees frequently miss work without reasonable excuses?
- Do some family members directly or almost directly supervise other family members?
- Do craftsmen complain that personal tools are stolen?
- Are company-owned small-tool losses high?
- Do small but valuable pieces of material or equipment "disappear" frequently?
- Is any completed or partially completed work mysteriously damaged?
- Have there been any fires of suspicious origin?
- Are any employees suspected of carrying concealed weapons?
- Have there been threats against any individual or group of individuals? Carried out on the job or off? Any unexplained cuts, bruises, black eyes?
- Are any employees evidently heavy drinkers or drug users? Do they exhibit symptoms of recent use of alcohol or drugs?
- Is there any evidence of drinking or drug use on the job (marijuana aroma, cigarette papers, empty beer cans)?
- Do employees or groups of employees seem nervous, jumpy, or otherwise unsettled?
- Do office supplies vanish quickly?
- Have personal items such as purses been stolen when no outside visitors have been in the office?
- Has management taken action to identify and remove the causes of theft, sabotage, drugs, and hostile behavior?
- Are underskilled employees identified and reclassified? Provided training?
- Are job rules clearly spelled out in the hiring process, and is comprehension acknowledged in writing?

- Is a formal labor relations-management-approved reprimand procedure in effect? Is the required documentation for each file or case kept safely?

ASSURING A SAFE WORKPLACE

- Are required OSHA records kept? Reports filed?
- Is emergency first aid available? Is an emergency medical evacuation plan in place? Have treatment arrangements been set up with medical practitioners and facilities?
- Do any employees hold current Red Cross first aid credentials?
- What is the current experience modifier or discount from manual or full rates for workmen's compensation premiums?
- Is housekeeping on the project site emphasized, left to chance, or used as a spare-time operation? Are trash receptacles convenient to drinking water, and are they promptly emptied when full?
- Are receptacles available for scrap and waste materials, and are they classified by material and emptied frequently?
- Are walkways strewn with electric cords, welding cables, rope, and scrap materials?
- Are electric tools, ladders, and other items requiring periodic OSHA inspection properly coded?
- Are employees issued safety belts when working above ground? Do they use them properly? Are ladders properly secured?
- Are new employees tested for safety knowledge and awareness before being put to work? Is proper safety gear issued?
- Are pending assignments carefully reviewed with craftsmen from an accident-prevention standpoint? Are safety meetings held frequently?
- Is recognition given for safety-goal achievement? Is the recognition meaningful? Is statistical information published?
- Does top job management visibly demonstrate its commitment to accident prevention? Frequently?

- Are safety records overemphasized? Are employees coerced into foregoing treatment?

BASIC PERSONAL COMFORTS

- Do most employee complaints or gripes center around basic personal comforts?
- Are basic personal comforts provided to everyone all the time?
- Are some basic personal comforts punitively withheld or eliminated?
- Is fresh, cold water provided at all work locations?
- Are field toilets clean, well-supplied, and well-maintained?
- Are toilets and drinking water routinely moved to new areas as the work progresses?
- Are adequate wash-up facilities provided?
- Are eating areas clean and protected from the weather?
- Are food- and drink-vending facilities available?
- Are trash barrels provided at all work locations?
- Is cleanup and trash removal performed regularly?
- Are parking areas crowded, dusty, and cluttered with debris?
- Are delays encountered coming to and leaving the site?
- Are parking areas secure? Are employees' vehicles damaged, vandalized, or burglarized?
- Are telephones available for use by craftsmen?
- Is rain gear issued when necessary?

TRAINING

- Is an on-the-job training program in place? Is it promoted?
- Have a representative number of employees been exposed to the program? What percentage complete it?
- Are off-job or after-hours training programs available?
- Is basic entry-skills training available?
- Is upgraded skills training available? Cross training?
- Is supervisory training available?
- Is recurrent training available? Updates on methods and techniques presented?
- Is training-session feedback passed on to management?

- Does management frequently drop in on training sessions?

RECOGNITION

- Are recognition programs for group and individual achievement in place?
- Do craftsmen receive praise for accomplishment?
- Are competition awards well-planned?
- Are promotion announcements well-planned?
- Are training certificate awards well-planned?
- Is recognition of achievement conveyed to craftsmen's families?
- Has there ever been a "Family Day" for tours of the project? More than one?
- Are recognition levels proportionate to the magnitude of accomplishment?
- Is a permanent system for tracking individual achievement in place? Are craftsmen aware of it?
- Are craftsmen's skills certified by the company? Are they given any visible proof such as a pocket card or wall certificate?
- Is any recognition given for group achievement?
- Are employees given direct cash incentives for production? Everyone or just a few? Open or confidential?
- Is recognition sometimes perceived as tainted by favoritism?
- Are perquisites controlled by management, or is management controlled by perquisites?

Index

MIX
Papier aus verantwortungsvollen Quellen
Paper from responsible sources
FSC® C105338

If you have any concerns about our products,
you can contact us on
ProductSafety@springernature.com

In case Publisher is established outside the EU,
the EU authorized representative is:
Springer Nature Customer Service Center GmbH
Europaplatz 3, 69115 Heidelberg, Germany

Printed by Libri Plureos GmbH
in Hamburg, Germany